To Sis & Paul,
Keepin' the music alive!
Chris

Authenticity in North America

This interdisciplinary book addresses the highly relevant debates about authenticity in North America, providing a contemporary re-examination of American culture, tourism and commodification of place.

Blending social sciences and humanities research skills, it formulates an examination of the geography of authenticity in North America, and brings together studies of both rurality and urbanity across the country, exposing the many commonalities of these different landscapes. Relph stated that nostalgic places are inauthentic, yet within this work several chapters explore how festivals and visitor attractions, which cultivate place heritage appeal, are authenticated by tourists and communities, creating a shared sense of belonging. In a world of hyperreal simulacra, post-truth and fake news, this book bucks the trend by demonstrating that authenticity can be found everywhere: in a mouthful of food, in a few bars of a Beach Boys song, in a statue of a troll, in a diffuse magical atmosphere, in the weirdness of the un-gentrified streets.

Written by a range of leading experts, this book offers a contemporary view of American authenticity, tourism, identity and culture. It will be of great interest to upper-level students, researchers and academics in Tourism, Geography, History, Cultural Studies, American Studies and Film Studies.

Dr Jane Lovell worked at the Royal Opera House Covent Garden and at Canterbury City Council, where she staged events including sculpture and international light shows. At Canterbury Christ Church University, Jane specialises in teaches Heritage and Creative Industry management and Creative Places. A cultural geographer, visiting fellow at the British Library and Associate Fellow at the UCL Institute of the Americas, she explores tourism, authenticity and places, magical spaces, film locations and researches the light installations that she continues to stage.

Dr Sam Hitchmough is Senior Lecturer in American Indian History at the University of Bristol, where he is also Director of Teaching/Programme Director in the History Department. He was previously Programme Director for American Studies at Canterbury Christ Church University. His research interests include the Red Power movement, the intersection between patriotism, protest and national narratives, Buffalo Bill's Wild West Shows in the UK, and the use of American Indian imagery in British popular culture.

Contemporary Geographies of Leisure, Tourism and Mobility
Series Editor: C. Michael Hall
Professor at the Department of Management, College of Business and Economics, University of Canterbury, Christchurch, New Zealand

The aim of this series is to explore and communicate the intersections and relationships between leisure, tourism and human mobility within the social sciences.

It will incorporate both traditional and new perspectives on leisure and tourism from contemporary geography, e.g. notions of identity, representation and culture, while also providing for perspectives from cognate areas such as anthropology, cultural studies, gastronomy and food studies, marketing, policy studies and political economy, regional and urban planning, and sociology, within the development of an integrated field of leisure and tourism studies.

Also, increasingly, tourism and leisure are regarded as steps in a continuum of human mobility. Inclusion of mobility in the series offers the prospect to examine the relationship between tourism and migration, the sojourner, educational travel, and second home and retirement travel phenomena.

The series comprises two strands:

Contemporary Geographies of Leisure, Tourism and Mobility aims to address the needs of students and academics, and the titles will be published in hardback and paperback. Titles include:

Overtourism
Tourism Management and Solutions
Edited by Harald Pechlaner, Elisa Innerhofer and Greta Erschbamer

Tourism Fictions, Simulacra and Virtualities
Edited by Maria Gravari-Barbas, Nelson Graburn and Jean-François Staszak

Community-Based Tourism in the Developing World
Community Learning, Development & Enterprise
Edited by Alan Clarke and Peter Wiltshier

Authenticity in North America
Place, Tourism, Heritage, Culture and the Popular Imagination
Edited by Jane Lovell and Sam Hitchmough

For more information about this series, please visit: www.routledge.com/Contemporary-Geographies-of-Leisure-Tourism-and-Mobility/book-series/SE0522

Authenticity in North America
Place, Tourism, Heritage, Culture
and the Popular Imagination

**Edited by Jane Lovell
and Sam Hitchmough**

LONDON AND NEW YORK

First published 2020
by Routledge
2 Park Square, Milton Park, Abingdon, Oxon OX14 4RN

and by Routledge
52 Vanderbilt Avenue, New York, NY 10017

Routledge is an imprint of the Taylor & Francis Group, an informa business

© 2020 selection and editorial matter, Jane Lovell and Sam Hitchmough; individual chapters, the contributors

The right of Jane Lovell and Sam Hitchmough to be identified as the authors of the editorial material, and of the authors for their individual chapters, has been asserted in accordance with sections 77 and 78 of the Copyright, Designs and Patents Act 1988.

All rights reserved. No part of this book may be reprinted or reproduced or utilised in any form or by any electronic, mechanical, or other means, now known or hereafter invented, including photocopying and recording, or in any information storage or retrieval system, without permission in writing from the publishers.

Trademark notice: Product or corporate names may be trademarks or registered trademarks, and are used only for identification and explanation without intent to infringe.

British Library Cataloguing-in-Publication Data
A catalogue record for this book is available from the British Library

Library of Congress Cataloging-in-Publication Data
Names: Lovell, Jane, editor. | Hitchmough, Sam, editor.
Title: Authenticity in North America : place, tourism, heritage, culture and the popular imagination / edited by Jane Lovell and Sam Hitchmough.
Description: London ; New York, NY : Routledge/Taylor & Francis Group, 2020. | Series: Contemporary geographies of leisure, tourism and mobility | Includes bibliographical references and index.
Identifiers: LCCN 2019039165 (print) | LCCN 2019039166 (ebook) | ISBN 9781138341319 (hbk) | ISBN 9780429440212 (ebk)
Subjects: LCSH: Popular culture—United States. | Popular culture—Canada. | Culture and tourism—United States. | Culture and tourism—Canada. | Authenticity (Philosophy)—Social aspects—United States. | Authenticity (Philosophy)—Social aspects—Canada.
Classification: LCC E169.Z83 A87 2020 (print) | LCC E169.Z83 (ebook) | DDC 306.0973—dc23
LC record available at https://lccn.loc.gov/2019039165
LC ebook record available at https://lccn.loc.gov/2019039166

ISBN: 978-1-138-34131-9 (hbk)
ISBN: 978-0-429-44021-2 (ebk)

Typeset in Times New Roman
by Apex CoVantage, LLC

To dear mum and dad. Thank you for having the confidence in me at the age of twenty to spend a wonderful year in America, studying in Albuquerque and eating sweet rolls in the Frontier restaurant.

To Leanne and Niamh – impossible without you. With love and thanks.

Contents

List of figures ix
Notes on contributors xi

Introduction – Hyper-authenticity 1
JANE LOVELL AND SAM HITCHMOUGH

1 **The kept weird: US American weird fiction and cities** 8
MC McGRADY

2 **'Something Like a Circus or a Sewer': the thrill and threat of New York City in American culture** 19
KEITH McDONALD AND WAYNE JOHNSON

3 **"That Chinese guy is where you go if you want egg foo yung": construction and subversion of exotic culinary authenticity in David Wong Louie's *The Barbarians are Coming*** 30
JIACHEN ZHANG

4 **Good authentic vibrations: the Beach Boys, California, and *Pet Sounds*** 41
CHRISTOPHER KIRKEY

5 **A Western skyline I swear I can see: affective critical rurality expressed through contemporary Americana music** 61
KEITH HALFACREE

6 **'We Sure Didn't Know': Laura Gilpin, Mary Ann Nakai, and Cold War politics on the Navajo Nation** 75
LOUISE SIDDONS

7 Opening the memory boxes: magical hyperreality, authenticity and the Haida people 96
JANE LOVELL

8 The authenticity paradox and the Western 112
KEN FOX

9 Playing at Westworld: gunfighters and saloon girls at the Tombstone Helldorado Festival 123
WARWICK FROST AND JENNIFER LAING

10 Hidden in the mountains: celebrating Swedish heritage in rural Pennsylvania 134
KATHERINE BURLINGAME

11 The triumph of trolls: the making, remaking and commercialization of heritage identity 145
ANN SMART MARTIN, CORTNEY ANDERSON KRAMER AND JARED L. SCHMIDT

12 'It is yet too soon to write the history of the Revolution': fashioning the memory of Thomas Paine 161
KRYSTEN E. BLACKSTONE

13 Familiarity breeds content: shaping the nostalgic drift in postbellum plantation life-writing 180
DAVID ANDERSON

14 Only going one way? *Due South*'s role in sustaining Canadian television 192
LINDA KNOWLES

Index 205

Figures

6.1 Laura Gilpin (1891–1979), *Francis Nakai and Family*, Sept. 17, 1950, gelatin silver print, 10 5/8 × 7 13/16 in., Amon Carter Museum of American Art, Fort Worth, Texas, Bequest of the artist, © 1979 Amon Carter Museum of American Art, P1979.128.509. 76

6.2 Unidentified Navajo artist, *Phase I Chief Blanket*, 1800–1850, Museum of Indian Arts & Culture/Laboratory of Anthropology, 9117/05. 85

6.3 Unidentified Navajo artist, *Phase I variation Chief Blanket*, 1850–1865, Museum of Indian Arts & Culture/Laboratory of Anthropology, 9118/01. 86

6.4 Laura Gilpin (1891–1979), *Mrs. Francis Nakai and Son*, 1932, gelatin silver print, 8 1/2 × 7 1/16 in., Amon Carter Museum of American Art, Fort Worth, Texas, Bequest of the artist, © 1979 Amon Carter Museum of American Art, P1979.128.82. 88

6.5 Laura Gilpin (1891–1979), *[Mrs. Francis Nakai]*, 1932, gelatin silver print, 9 11/16 × 7 3/4 in., Amon Carter Museum of American Art, Fort Worth, Texas, Bequest of the artist, © 1979 Amon Carter Museum of American Art, P1979.128.711. 89

9.1 Westworld comes to Tombstone: re-enacting a gunfight (source Warwick Frost). 128

11.1 Trolls greeting visitors to the Open House Import Store on Main Street in Mt. Horeb. Wood. Image by Jared L. Schmidt. (Included in chapter). 146

11.2 Norwegian immigrant trunk. MHAHS 2014.073.0020. Pine case, iron hardware. The maker is unknown, but the trunk was painted by Ole Haugen, c. 1835. From the original Little Norway Collection acquired by Mt. Horeb Area Historical Society. Image courtesy of the Mount Horeb Area Historical Society, J. Bastien, photographer. 151

11.3 Cupboard MHAHS 1984.011.0001. Pine, metal, glass. Made by Aslak Olson Lie c. 1860. Image courtesy of the Mount Horeb Area Historical Society, J. Bastien, photographer. 154

11.4	Dress in a modern A-line cut featuring stitched rosemaling. MHAHS. 1990.014.0003 Made by Oljanna Cunneen c. late 1960s. Wool. Gift from its owner, Nancy Vogel. Image courtesy Mount Horeb Area Historical Society. J Bastien, photographer.	156
11.5	A&W Root beer and Restaurant company troll sign. MHAHS 1999.098.0014. Plywood. Painted by Patricia Edmundson, c. 1984. Image courtesy Mount Horeb Area Historical Society, J. Bastien, photographer.	157
12.1	This is the Paine statue in New Rochelle as it appears in 2011. In 1839 when the monument was originally unveiled, it would not have had the bust on the top, or inscriptions on the sides. Copyright: Peter Radunzel, 'Thomas Paine Memorial', (2011).	166
12.2	The statue of Thomas Paine in Burnham Park Copyright: Dan Beards, 'Thomas Paine in Burnham Park, Morristown N.J', (2008).	171
12.3	Paine with writings, with a gun at his feet, highlighting service to and in the Continental Army during the Revolution as a writer and Aide de Camp. Copyright: Lee B. Spitzer, 'Bordentown Thomas Paine Statue', (2017).	174

Contributors

David Anderson is Senior Lecturer in American Studies at Swansea University in Wales, where he has worked since 2008. He has published in the Journal of Southern History, Civil War History, and Southern Studies and is currently completing a monograph on the Lost Cause and American Civil War memory.

Cortney Anderson-Kramer is an Art History PhD candidate at the University of Wisconsin-Madison, primarily interested in American Folk and Outsider art. Anderson-Kramer and Schmidt were student contributors to the Mount Horeb Historium's inaugural exhibition under the direction of Professor Martin.

Krysten E. Blackstone is a History PhD student at the University of Edinburgh, where her research examines morale and identity in the Continental Army during the American Revolution. Additionally, she holds research fellowships at the Society of the Cincinnati, the British Library and the Fred W. Smith National Library for the Study of George Washington.

Katherine Burlingame is a PhD candidate in the Department of Human Geography at Lund University. Her current research focuses on heritage management and tourism development issues in rural landscapes. More broadly, Katherine's research interests include intangible heritage, phenomenology, storytelling, belonging and creative methods.

Ken Fox teaches film in the School of Creative Arts and Industries, Canterbury Christ Church University. He is School Director of Learning and Teaching, a University Teaching Fellow and a Senior Fellow of the Higher Education Academy. Ken teaches and researches about the representation of landscape and cityscape in cinema and television. Ken is a native of Donegal in the Republic of Ireland where he returns with his family most summers to experience the authentic/inauthentic experience of being a tourist in his own country.

Warwick Frost is Professor in the Department of Management, Sport and Tourism at La Trobe University, Australia. His research interests are in environmental history, frontier history and representation, cultural heritage, events and tourism, and the media. He is a convenor of the biennial International Tourism and Media conference series.

Keith Halfacree is Reader in Human Geography in the Department of Geography at Swansea University in the UK. His research focuses on four overlapping areas. First, Keith studies mostly "internal" expressions of human migration and has published widely on urban-to-rural "counterurbanisation," cultural perspectives on migration, and on how migration helps structure institutions such as family and place. Second, he investigates discourses of rurality, mostly in the global North, outlining the content of our representations of the rural and exploring how these link to practices such as migration, developing a sense of place and leisure. Third, building on the latter, he engages with conceptual debates on rural change and the shaping of rural futures. Finally, and very much reflecting his personal interests, Keith teaches and researches "marginal geographies," or geographical expressions of all things "counter-cultural," notably here including a strong focus on critical aspects Americana and other forms of popular music.

Wayne Johnson is Senior Lecturer in Media and Film Studies at York St. John University. He received his PhD from Keele University. He is currently working on a book on contemporary horror and his current areas of research are in film and political and social change, the contemporary Western, and American culture and social history.

Christopher Kirkey is Director of the Center for the Study of Canada and Institute on Quebec Studies at State University of New York College at Plattsburgh, New York. A scholar of comparative foreign policy and international relations theory, he has been a professor at Bridgewater State College (1993–2001), Columbia University (2002–2012) and SUNY Plattsburgh (2002–present). His recent works include *Canadian Foreign Policy in a Unipolar World* (Oxford), *Quebec Questions: Quebec Studies in the Twenty-First Century* (Oxford), "The Quebec Election of April 2014: Initial Impressions," (*London Journal of Canadian Studies*), various co-edited journal issues including the *Journal of Eastern Townships Studies*, *American Review of Canadian Studies*, *Québec Studies* and the *British Journal of Canadian Studies*. He is currently working on nine scholarly projects, including books, co-edited books, book chapters and upcoming special journal issues. Dr Kirkey serves on the editorial board of the *American Review of Canadian Studies*, the *International Journal of Canadian Studies* and the *London Journal of Canadian Studies*. He is co-editor of the *Journal of Eastern Townships Studies* and is President-Elect of the Association for Canadian Studies in the United States (ACSUS).

Linda Knowles is a Canadian independent scholar who has lived in the United Kingdom for the past forty years. She is a member of the British Association for Canadian Studies.

Sam Hitchmough is Senior Lecturer in American Indian History at the University of Bristol where he is also Director of Teaching/Programme Director in the History Department. He was previously Programme Director for American Studies at Canterbury Christ Church University. His research interests include the Red Power movement, the intersection between patriotism, protest and national narratives, Buffalo Bill's Wild West Shows in the UK, and the use of American Indian imagery in British popular culture.

Jennifer Laing is Associate Professor in the Department of Management, Sport and Tourism at La Trobe University, Australia. Her research interests are in travel narratives, the role of events in society, health and wellness tourism and tourism and the media. She is a convenor of the biennial International Tourism and Media conference series. In 2017, she was named as the Emerging Scholar of Distinction by the International Academy for the Study of Tourism.

Jane Lovell worked at the Royal Opera House Covent Garden and at Canterbury City Council, where she staged events including sculpture and international light shows. As Senior Lecturer at Canterbury Christ Church University, Jane teaches Heritage and Creative Industry Management and Creative Places. She is a Council Member of the British Association of Canadian Studies, a Visiting Fellow at the British Library and an Associate Fellow at the UCL Institute of the Americas. A cultural geographer, she explores tourism, authenticity and places, in particular American frontier, magical spaces, film locations and researches the light installations that she continues to stage.

Ann Smart Martin is the Stanley and Polly Stone (Chipstone) Professor in the Art History Department at the University of Wisconsin-Madison, where she teaches material culture, decorative arts, vernacular arts and exhibition practice, She directs the university's interdisciplinary material culture program, has published widely and curated ten exhibitions, including the one on which this research is based.

Keith McDonald holds a PhD from Birkbeck College, University of London and is a Senior Lecturer in Media and Film and a Subject Director at York St John University. His is the author of *Guillermo del Toro: Film as Alchemic Art* (Bloomsbury, 2014) and is currently co-writing a book on contemporary Gothic film for Anthem Press. His research interests include popular fiction, new media and pedagogy.

MC McGrady is a theorist and independent game designer. They received their BA from UC Santa Cruz, their MA from the University of Chicago and are completing their PhD at Cardiff University, where they are researching the links between "nerd" culture and the resurgence of fascism in the United States.

Jared L. Schmidt is a PhD student at the University of Wisconsin-Madison enrolled in the Department of Comparative Literature and Folklore Studies. He also holds a master's degree in Applied Anthropology from Minnesota State University-Mankato. His research interests include heritage, material culture, museums, foodways and digital folklore.

Louise Siddons (PhD, Stanford University) is Associate Professor of Art History, teaching courses in American and Native American visual culture at Oklahoma

State University. She has published on topics from the eighteenth century to the present, and is active as an independent curator and critic. Siddons was the founding curator and co-director of the Oklahoma State University Museum of Art; her most recent monograph is *Centering Modernism: J. Jay McVicker and Postwar American Art* (University of Oklahoma Press, 2018). She is currently writing a book about photographer Laura Gilpin and midcentury Navajo sovereignty.

Jiachen Zhang is Lecturer in the School of Foreign Languages and Literature, Fudan University, China. He just obtained his PhD from the School of English, the University of Leeds. His doctoral research investigates representations of food and abjection in Asian American fictions published between 1989 and 2003. His articles appear in *Ethnicity and Gender Debates* (Berlin: Peter Lang, forthcoming) and *Offal: Rejected and Reclaimed Food* (Oxford: Prospect Book, 2017). He also writes book reviews for US Studies Online.

Introduction – Hyper-authenticity

Jane Lovell and Sam Hitchmough

The "real" North America explodes kaleidoscopically into a myriad of meanings and narratives. There are multiple volumes that discuss aspects of the authenticity of the American Dream hyperreal simulacrum of Disney, Disneyisation (Baudrillard, 1986; Eco, 1986; Bryman, 2004) or the frontier myth (Stoeltje, 1987; Slotkin, 1992; Spurgeon, 2005), and a traditional and growing body of work that seeks to theorise authenticity, particularly with regard to tourism (Boorstin, 1961; Cohen, 1988; MacCannell, 1976; Vidon *et al.*, 2018) and heritage (Labadi, 2010), but the emergent notions of authenticity's criticality in understanding "America" have been overlooked. This book examines authenticity by observing its different constructions, all of which highlight the fact that North America is not one place, instead it is composed of an assemblage of meanings.

Myths concern the origins of a people, often intertwined with a supernatural element, and their key characteristic is that they are continuously adapted (Barthes, 1957). America is composed of many myths, and some are regional—The West, the South, the Pacific Northwest. The book unpacks some of these myths, tracing their origin to the music, art, literature, philosophy, craftsmanship and forms of intangible cultural heritage which reside at the heart of the place. Sense of place is a construct and it is composed of fragments and traceries of these elements. Tourism and the study of places provides the ideal crucible in which to examine the authenticity of the ways in which host communities view themselves, the ways in which place marketing can shape community identity, and the direct, unscripted encounters between hosts and travelers providing a "backstage". The authenticity of place may be experiential. When we travel, it is often to an appropriate nostalgic soundtrack, while sampling a site-specific menu and trying local delicacies, for example, sitting in the Frontier Café in Albuquerque eating cinnamon sweet rolls while watching the mountains turn watermelon pink in the distance. We authenticate places by purchasing a volume by the Beat authors in the City Lights bookstore in San Francisco or posing in front of a historic statue for a selfie. In different eras we encounter symbols and chimeras of authenticity which draw us to the study of America, such as the urban graffiti of *Kavalier and Clay*, Spike Lee's Brooklyn, Sam Shephard's *True West* and *Silent Tongue*, Walter White's pork pie hat, Agnes Martin's minimal paintings or the cut-paper silhouettes of Kara Walker.

Fundamentally shaped by forces ranging from memory to political agendas, and from nostalgia to discrimination, place-making and identity have always been powerful processes in the US, and so frequently at the heart of it all is a notion of what constitutes the "real" America. Blowing off the desert dust, stripping off the corporate veneer, untangling the vertiginous skyscrapers—what lies beneath for individuals, groups, communities, organizations, politicians, musicians, artists and writers is a core set of values that infuse and knit together the idea of place and self. How do people conceive of American authenticity? How are these multiple authenticities created and how can they coexist? How do people experience authenticity? There is much excitement about engaging with the range of authentic locations in America, but much of this immersion revolves around staged authenticity or inauthenticity.

A formidable array of lenses distort and warp any leverage we might have on authenticity. Theme parks, historical heritage sites, national parks and even the "wilderness" all mimic and copy authenticity to the point that it inverts it. The lyrics of Bruce Springsteen's track "The River" ask, "is a dream a lie if it don't come true, or is it something worse", and the American Dream itself is in now question. American heritage is undergoing what can be described as an "inauthenticity turn" and, arguably and critically, this can be extended to North America. Baudrillard (1988) argued that the construct of reality becomes more real than the original, a hyper-authenticity. The fake news and alternative facts of this Trumpian era result in a questioning of what the real America is, what it stands for, and how it remembers and "curates" itself. This volume brings together an eclectic range of pieces that all explore versions of North American authenticity and inauthenticity as a fluid concept dependent on time and place. Authenticity can be something thought to be lost that needs to be retrieved, something that needs to be celebrated, commemorated or challenged, something that has been altered through erasure or imposition, something that is appropriated and "mainstreamed" that needs to be reclaimed, or something that has been co-opted, sanitized and readied for capitalistic consumption. Authenticity is simultaneously functional, cannibalized, transactional, "real", "fake", fictional, oppositional and hyper.

Above all, authenticity is, as a concept, intersectional. And herein lies its most dynamic use—its ambiguity, stickiness and slipperiness is echoed when we interrogate North America as a subject. The book brings together researchers from American and Canadian Studies, Tourism, Heritage, Film and Geography backgrounds to demonstrate that the Humanities and the Social Sciences share the same concerns. The creative or historical works discussed by Humanities form the subjects of interest for the Social Sciences. Above all, these chapters are all animated by a desire to tug at the frayed thread of the American tapestry and work out what was holding it together and what it is made of. In the mix of purposes and functions that authenticity serves, what does this allow us to understand about America as a nation and as an idea? By looking at how postbellum life-writers conjured an authentic sense of place and identity—how New York constantly shimmers as an illusion, how the idea of being authentic through "weirdness" can expose the inherent co-option and commercialization—case-studies allow us to discern important patterns. This process has history. Smoothing out difficult history and negotiating the past have been critical to holding onto important ideas of an authentic America for different groups of people.

To traverse these ideas, take a road trip with us to explore the ways in which we authenticate places and they authenticate us. First stop, off at Austin, Santa Cruz and Portland, where McGrady guides you through some forms of carefully constructed authenticity. Drawing connections between examples of the early literary weird and 21st century "keep weird" campaigns of McGrady argues that the "kept weird" of civic slogans should be considered a particular iteration of the wider phenomenon known as "the weird". Through a discussion that knits together civic violence, commodification, hipsterisation, gentrification and the capitalist co-option of what might have previously been considered radical or threatening, the kept weird is considered a transactional relationship wherein ostensibly fringe identities are allowed within the bounds of civic culture so long as they can be made safe, sanitised and readied for capitalist consumption. Next up, New York. Cities are presented by McDonald and Johnson as dynamic spaces in which glimpses of "authentic" Americanness interact with corporatism and spectacle in their discussion of New York City, a location and concept that represents the absolute essence of American authenticity whilst being, simultaneously, entirely fake. Applying Derrida's (2014) "hauntology", New York embodies a complex hyper-text, a mediated bricolage of contradictory and complimentary texts that have embedded themselves into the cultural imagination. Employing a wide palette including *The Walk* and *The Avengers*, they conclude with a consideration of the 2016 Times Square concert by Jay-Z and Alicia Keys, a supreme demonstrative example of space as "text"—a spectacle set against a city that is simultaneously fabricated and viscerally authentic.

Take time for dinner in the Richfield Ladies' Club. Place identity and its hybridisation is at the heart of Zhang's chapter exploring representations of exotic culinary authenticity in Chinese American writer David Wong Louie's debut novel, *The Barbarians are Coming* (2000). The chapter unpacks Lisa Heldke's (2003) discussion of the Euro-American fascination towards authentic exotic cuisines. In *Exotic Appetites*, Heldke (2003, pp. 23–44) suggests that the desire thrives on ideas of the essence or the purity of origins that are conventionally understood to be static and unchanging. In the face of the dominant culture's ongoing racial appropriation of the Asian body and culture, this chapter investigates how the protagonist constructs an alternative and politicised culinary authenticity; that by eating "real" Chinese food they are able to challenge US racial politics and imagine the recuperation of Chinese collective memories.

Leaving the cityscapes behind and turning up the dial, Kirkby's chapter explores the sound of authenticity in *Pet Sounds*, an album by The Beach Boys which was "tethered in the sun, sand and surf of Southern California" (Kirkby). Kirkby takes a granular approach to the subject matter, exploring the recording history, critical reception and impact of the *Pet Sounds* album. Kirkby suggests, "that *Pet Sounds* is the most musically advanced recording undertaken by the Beach Boys, and arguably represents the greatest authentic American rock record ever created".

The chapter investigates the authenticity of the musical arrangements, melodies, instruments utilized, chord progressions, vocal harmonies and lyrics of each of the thirteen songs; the reception given to *Pet Sounds* upon its release; and an assessment of its long-term impact and place in rock music history. Changing the radio station, music is also central to Keith Halfacree's analysis of authenticity in rural and small-town America, and how the voices of (a geographically extended) neglected rural America are expressed and represented through the Americana music of Willy Vlautin. In so doing, Halfacree contrasts the West's strongly positive symbolic imagery drawn from the still deeply embedded Frontier myth with the increasingly marginalised everyday lives of abandoned rural populations. Utilising the concept of affective critical rurality, he discusses how Vlautin articulates hardships but expresses them differently from either right-wing populism or Frontier Myth essentialism. Mobility, for example, is expressed as a response to conditions less as celebrating rugged individualist freedom but more frequently as misguided and ineffective. The best "freedom" many can hope to attain is through the modest existential celebration of small, emotionally resonant places and experiences.

Meanwhile, Siddons takes out the camera and focuses it. The chapter explores the Anglo photographer Laura Gilpin's 1950 photograph of a Navajo family (published in *The Enduring Navajo*, 1968) and how it queered conversations about authenticity and national identity that were taking place on the Navajo Nation during the Cold War. In *The Enduring Navajo*, Gilpin resisted government-imposed definitions of Navajo and American authenticity, and urged readers to shift their perspective on nationalism and indigenous sovereignty. Via a discussion of Gilpin's own change in perspective that saw a move away from an ethnographic, paternalistic and romanticising vision of Native American culture to a more contemporary understanding of Navajo politics, Siddons visually analyses photographs that effectively disrupt the complacency of settler colonialism, patriarchy and American nationalism.

Arriving by boat at the peripheral communities in the liminal landscape of the Pacific Northwest, Lovell's chapter investigates how perceptions of the "magically real" authenticity of Haida art and culture and the of the Haida Gwaii archipelago have emerged over time. "Place myths" can be generated in tourism (Shields, 1991) by interweaving the history of places with the symbolic to create storied landscapes, and the article examines how the "Super, Natural" branding of British Columbia forms a multilayered simulacrum. The study explores the presentation of an essentialist, magically real Haida Gwaii in tourist literature ranging from the World Heritage Site's international reach to national, regional, local and individual business strategies using a discourse of authenticity. The article asks about the true meaning of participatory planning and ownership. Having once been defined by ethnographers such as Boas as pre-modern, the Haida now emphasise their contemporary "living culture" in local strategies that focus on the plural dialogue of "language, art, stories and voices" at the community hub of the Haida Heritage Centre. The study suggests that there is still an underlying colonial strategic emphasis on the old world as a commodity of the new; however, meaningful interpretative encounters, heritage stewardship and performances to

Introduction – Hyper-authenticity 5

a relatively small numbers of visitors to Haida Gwaii shifts the power. Simultaneously, images of Haida art circulate and are copied online, creating a debate about dispossession, appropriation, and the meaning of authenticity and decolonialism in the age of digitisation and hyperreality. The chapter problematises how, when place-myths are elevated to a regional and arguably international level, they become simulacra—"magically hyperreal".

Out West, Fox's chapter explores the tropes of the myth of the West, asking, if, as Baudrillard (1994) suggests, "illusion is no longer possible, because the real is no longer possible", why place the terms authenticity and Western genre in the same sentence? How can authenticity be claimed for the Western genre when in Baudrillard (1988) terms, it is its own simulacrum? Fox argues that this constitutes an "authenticity paradox" which is complicated and multilayered because the Westerns are "spliced" with other genres to form hybrid productions of films depicting the West. Focusing on two films—*Hostiles* and *The Lone Ranger*—with contrasting narrative intentions, both drawing from the genre's bank of iconographic images, the chapter unfolds the awe, wonder and spectacle of American sublime landscape as place and metaphor. Staying in the West, Frost and Laing's' chapter concerns The Tombstone Helldorado Festival, which, as it suggests is, "a carnivalesque and entertaining mélange". It raises important central questions regarding the transformative authenticity of self, with participants dressing up and acting as historical characters. As the authors argue, "Such events might be fun, but they also have something to say about the society that creates them". The re-enactments at Tombstone emphasise authenticity, in the "conspicuous display of Wild West costumes, weaponry and other paraphernalia", while "there is almost no interest in recreating historical stories". Instead, the displays draw on the mythology of the West using cinematic motifs. As the authors contend, authenticity is vested in the "spirit" rather than the minutiae. Yet the influence of the movies leads to violence towards women and issues of race that mirror the HBO series *Westworld*. The chapter speaks of the authenticity of the identity of reenactment participants who find in the past a role they no longer have in society. Yet also a past that is whitewashed with marginal groups strongly underplayed, "with little or no evidence of desire to undertake Black, Hispanic, Chinese or Native American roles".

Moving back East, another chapter speaks of the search for authenticity and its expression through festivals. Burlingame's chapter articulates issues related to the authenticity of time and space through residents' desire to belong. The festival celebrates a past that "not only resonates with individuals searching for rootedness, but also with rural communities whose unique histories become obscured with time". As younger residents move away, older people within the community foster the past through festivals, such as the Mt. Jewett Swedish Festival in the state of Pennsylvania. The chapter explores the experiential authenticity invested in the sense of nostalgia, "belonging" and identity, as a resident stated, "The festival allows for the opportunity to celebrate the real and 'imagined' community". The rituals and traditions involved in this festival are deep-seated. Another expression of "real and imagined" authenticity is approached in a chapter tracing the creation of the The Driftless Historium and History Center in Mount Horeb, a volunteer-founded organisation that amassed records and heirlooms reflecting the European ethnic roots of the community. In parallel, the chapter investigates how Mount Horeb also "branded" itself as "Norwegian place" by creating a "Trollway" through the center of town. The Trollway features thirty-seven wooden sculptures of trolls, the creature of Norweigan folklore. They are carved by chainsaw, dressed and styled to be uncanny. They themselves are now "vernacular landmarks", giving Mount Horeb cultural heritage distinctiveness as a "concept town" (Gradén, 2003) themed to enhance the rural tourist economy. The work assesses the development and marketing of heritage tourism, in particular the objects and landscapes themselves, which shape authenticity of place and identity.

We stay eastward for Blackstone's chapter that considers the authenticity of memorialization and constructions of social memory embodied in four statues of Thomas Paine erected in the United States in 1839, 1881, 1952 and 1997. Each statue reflects the emergent authenticity (Cohen, 1988) of different considerations; in the nineteenth century, broad appropriation of Paine, even amongst otherwise conflicting factions, established the pattern of diversity and controversy over Paine's memory that subsequent generations followed. However, in the twentieth century, to become nationally acceptable, groups modified the memory of Paine to include only his accomplishments connected to the Revolution. This ignored his more controversial traits and made his memory more appropriate for popular consumption. Generations of radicals and reformers maintained Paine's memory, each utilising it in a unique way. The four monuments to Paine that exist in the United States each propagate a different version of him and are highly affected by the society surrounding their creation. The ebb and flow of regard for Paine since his death demonstrates that memories do not regulate the present; the present dictates what memories get recalled to suit its purposes (Kammen, 1991, p. 13). However, there seems to be an innate authenticity to Paine's words, which, like President Obama claimed, have become "timeless".

Down South, nostalgia is central to the way in which Anderson unpacks how many white southerners in the postbellum period attempted to reframe the idea of the South. It employs the hitherto largely under-explored autobiographies and memoirs of former plantation owners to explore how memories of those writing in the postbellum period were filtered and reframed to offset circumstances of change and uncertainty. Using relatively overlooked sources, it quickly becomes clear how this group chose to remember and reimagine their recent past. The Civil War was cast as a rift in the fabric of time, and is an event that created a sense of historical discontinuity and conjured an uneasiness around the idea that the remembered past was receding. It created two different worlds, a pre- and post-war South that needed to somehow be negotiated to reconstruct continuities and familiar patterns. Nostalgia helped to define "their" South, and the author. The nostalgia that evoked notions of close affinities between paternal master and loyal retainers, Anderson reveals how postbellum life-writing became a rhetorical strategy designed by whites to project an authenticity onto ideas of identity and outline a future for southern race relations that adhered to the hierarchical patterns of slavery, illuminating a different authenticity.

We end our journey in Chicago, turning on the motel TV for a nostalgic look at Knowles' chapter. In 1994, *Due South* became the first Canadian-produced series to air in primetime on a major US television network. Created by Paul Haggis, the series follows the adventures of a Royal Canadian Mounted Police Constable, Benton Fraser, who "first came to Chicago on the trail of the killers of [his] father and, for reasons which don't need exploring at this juncture, . . . remained, attached as liaison to the Canadian consulate." This chapter examines *Due South* in the context of Canadian television's unequal struggle with the American industry and the consequent masking of Canadian identity in homegrown productions. It also explores the Canadian tendency to promote the myth of the northern wilderness over the urban reality of most Canadians.

The road trip will never end. The one certainty of myth is that it always adapts; to research authenticity is to travel towards an ever-retreating destination.

References

Barthes, R. (1972, first published 1957) *Mythologies*. Translated by Annette Lavers. New York: Hill and Wang.
Baudrillard, J. (1986) *In America*. London: Verso.
Baudrillard, J. (1988) *Selected Writings: Simulacra and Simulation*. Cambridge: Polity Press.
Baudrillard, J. (1994) *Simulacra and Simulation*. Translated by Sheila Faria Glaser. Ann Arbor: University of Michigan Press.
Boorstin, D. (1961 and 1964) *The Image: A Guide to Pseudo-Events in America*. New York: Harper & Row.
Bryman, A. (2004) *The Disneyization of Society*. London: Sage.
Cohen, E. (1988) 'Authenticity and Commoditization in Tourism', *Annals of Tourism Research*, 15, pp. 371–386.
Derrida, J. (2014) *Specters of Marx: The State of the Debt, the Work of Mourning and the New International*. Abingdon: Routledge.
Eco, U. (1986) *Faith in Fakes: Travels in Hyperreality*. London: Picador.
Gradén, L. (2003) *On Parade: Making Heritage in Lindsborg, Kansas*. Upsaliensis: Rock Island.
Heldke, L. (2003) *Exotic Appetites: Ruminations of a Food Adventurer*. London: Routledge.
Kammen, M. (1991) *Season of Youth: The American Revolution in Historical Imagination*. New York: Alfred Knopf.
Labadi, S. (2010) 'World Heritage Authenticity and Post Authenticity: National and International Perspectives', in Labadi, S. and Long, C. (eds.) *Heritage and Globalisation*. Abingdon: Routledge, pp. 66–84.
MacCannell, D. (1976) *The Tourist: A New Theory of the Leisure Class*. New York: Solouker Books.
Slotkin, R. (1992) *Gunfighter Nation: The Myth of the Frontier in Twentieth-Century America*. Norman, Oklahoma: University of Oklahoma Press.
Spurgeon, S. (2005) *Exploding the Frontier: Myths of the Postmodern Frontier*. College Station, Texas: Texas A&M University Press.
Stoeltje, B. J. (1987) 'Making the Frontier Myth: Folklore Process in a Modern Nation', *Western Folklore*, 46(4), pp. 235–253.
Vidon, E. S., Rickly, J. M. and Knudsen, D. C. (2018) 'Wilderness State of Mind: Expanding Authenticity', *Annals of Tourism Research*, 73, pp. 62–70.

1 The kept weird

US American weird fiction and cities

MC McGrady

USAmerican weird fiction

This essay is best read as a kind of culture and travel guide to important destinations in the psychopolitical landscape of the United States, as the cities of Austin, Santa Cruz, and Portland are tied together by shared patterns of cultural materiality that seem inexorably to erupt in reactionary violence. While 'the weird' as such and the 'kept' variety of the title have specific textual origins – the former appearing as national and transnational instances of genre fiction in the early 20th century (VanderMeer and VanderMeer, 2012, pp. 5–6) and the latter in the shared slogan 'Keep [City] Weird' that emerged in the first years of the 21st (Yardley, 2002; Hoppin, 2013) – the phenomena under discussion here are larger, embodied by the intersectional conflicts arising from the violent maintenance of imperial white-supremacist capitalist patriarchy by state and para-state forces, leftist resistance to this violence and the liberal bourgeois neutralization of this resistance, all filtered through a national identity written alongside the expansion of the country's borders (first through settler colonialism and then a military empire). With all that in mind, the easiest place to start is at the country's textual inauguration in the form of the manifesto known as the Declaration of Independence.

The early contours of the USAmerican weird ('USAmerican' is used here in lieu of the more common metonymic use of 'American' in recognition of the degree to which the latter is part and parcel of the United States' imperial actions in South America) can be found in the Declaration's list of 'repeated injuries and usurpations' committed by England's King George III (US 1776). When the Declaration accuses the King of dissolving 'Representatives Houses repeatedly, for opposing with manly firmness his invasions on the rights of the people', it not only adopts the still-common patriarchal metonymic use of 'people' to mean 'white male land owners', but with the term 'manly firmness' prefigures the more euphemistic 'broad shoulders' preferred by contemporary white male USAmerican politicians when discussing geopolitics (Ross, 2016). With this, the ontological center of what would become the United States is marked out – white men whose claim to a gendered power is already defined by the anxious need to exclaim that power.

Against this is not only King George, but also 'the inhabitants of our frontiers, the merciless Indian Savages, whose known rule of warfare, is an undistinguished destruction of all ages, sexes and conditions'. With this, the basic features of the USAmerican weird are set, as the privileged position of white men is threatened by an inhuman other that defies epistemological limits and emerges from outside the bounds of civilization or meaning.

Two examples from the primordial stages of weird fiction make clear how these themes developed alongside the expanding USAmerican frontier, at least until the genocide of indigenous people in the United States shifted from expansionary military operations to 'internal' police functions. The first is Nathaniel Hawthorne's short story 'Young Goodman Brown', published in the April 1835 issue of *The New-England Magazine* but set in 17th-century Salem Village as a young Puritan journeys to meet a mysterious figure in the forest at night. As Goodman Brown takes 'a dreary road, darkened by all the gloomiest trees of the forest', the landscape outside the village is immediately tied to a teeming, unknowable threat:

> It was all as lonely as could be ; and there is this peculiarity in such a solitude, that the traveller knows not who may be concealed by the innumerable trunks and the thick boughs overhead ; so that, with lonely footsteps, he may yet be passing through an unseen multitude. 'There may be a devilish Indian behind every tree,' said goodman Brown, to himself; and he glanced fearfully behind him, as he added, 'What if the devil himself should be at my very elbow!'
> (Hawthorne, 1835, p. 250)

When the devil does appear in the form of Brown's travelling companion, he announces his own participation in the expansion of the frontier, remarking that

> I have been as well acquainted with your family as with ever a one among the Puritans; and that's no trifle to say. I helped your grandfather, the constable, when he lashed the Quaker woman so smartly through the streets of Salem. And it was I that brought your father a pitch-pine knot, kindled at my own hearth, to set fire to an Indian village, in king Philip's war.
> (Hawthorne, 1835, p. 251)

Here, the devil does not refute Brown's impression that the forest holds some unknowable horror, but rather points out that the horror is Brown's own complicity in colonization, rather than the indigenous victims of that intrusion. The devil goes on to reveal his association with all of New England, claiming that

> [t]he deacons of many a church have drunk the communion wine with me; the selectmen, of divers town, make me their chairman; and a majority of the Great and General Court are firm supporters of my interests. The governor and I, too – but these are state-secrets.
> (Hawthorne, 1835, p. 251)

One can easily extend these associations to the United States more widely by noting the specific act the devil accuses Brown's father of, as both the first USAmerican president George Washington and his great-grandfather John Washington were given the nickname 'Conotocarious' by the Iroquois, which translates to 'town taker', 'burner of towns', or 'devourer of villages' (George Washington's Mount Vernon, 2018).

The second example is Ambrose Bierce's, 1897 short story 'The Eyes of the Panther', which looks back favorably at the early days of USAmerican colonization even as it expresses the colonial anxieties of miscegenation and racialized cuckoldry in the form of a panther. The equivalent of Hawthorne's Goodman Brown is Bierce's Charles Marlowe, who is

> of the class, now extinct in this country, of woodmen pioneers – men who found their most acceptable surroundings in sylvan solitudes that stretched along the eastern slope of the Mississippi Valley, from the Great Lakes to the Gulf of Mexico. For more than a hundred years these men pushed ever westward, generation after generation, with rifle and ax, reclaiming from Nature and her savage children here and there an isolated acreage for the plow, no sooner reclaimed than surrendered to their less venturesome but more thrifty successors.
>
> (Bierce, 1897)

Here, the effects of colonization on the landscape are clear, as the relationship of white men to the forest and its 'savage children' reflects the expansionary assumptions of Manifest Destiny; where once Goodman Brown journeyed into a forest beyond the village-bound limits of white male hegemony, in Bierce's story, colonizers are *reclaiming* a landscape that has been usurped by Nature and its indigenous inhabitants. To position Nature as an unnatural imposition on white men's 'acreage' reflects the oxymoronic logic of white settler identity, which must frame the resistance to invasion as an initializing violence rather than a response to the initial violence of invasion.

The racialized threat of sexual violence and cuckoldry appears in the form of a panther that terrorizes Marlowe's wife and child (unnamed except for the titles 'the wife' and 'Baby'). In the story, the panther never attacks but rather keeps Marlowe's wife cowering 'in absolute silence', with 'the moments growing to hours, to years, to ages' as she gradually smothers her baby (Bierce, 1897). Three months after the event she dies in childbirth, though this child – named Irene – grows up to be 'young, blonde, graceful' with eyes described as 'gray-green, long and narrow, with an expression defying analysis. One could only know that they were disquieting. Cleopatra may have had such eyes' (Bierce, 1897). Despite the three-month gestation that complicates a literal reading, the panther in the story cannot help but appear as an allusion to the perceived threat of miscegenation and racialized sexual violence, particularly as the story's pre-Civil War setting allows for the possibility that the panther represents an even more complex racialized threat in the form of the maroon communities made up of runaway slaves and indigenous people. This fear eventually leads to Irene's death, as a rejected suitor shoots her in the night when he apparently mistakes her for a panther, making clear the degree to which white men's fear of racialized sexual violence is directly tied to their own violence against women.

In the 2016 book *The Weird and the Eerie*, cultural critic Mark Fisher rejects the notion that Hawthorne or Bierce might be considered part of the weird, comparing them unfavorably in this regard to H.P. Lovecraft and writing that 'any discussion of weird fiction must begin with Lovecraft', who wrote for the pulp magazine *Weird Tales* (pp. 16, 18). Instead, Fisher considers Hawthorne and Bierce to be 'Gothic novelists' whose work lacks 'Lovecraft's emphasis on the materiality of the anomalous entities in his stories' (Fisher, 2016, p. 18). Fisher's expulsion of Hawthorne and Bierce from the catalogue of weird fiction does not by any means indicate a widely accepted boundary for the genre (Bierce, 2018), but the way he focuses on Lovecraft is instructive for appreciating how thoroughly the anxieties that permeate 'Young Goodman Brown' and 'The Eyes of the Panther' characterize the USAmerican weird.

Although Fisher points to 'the supreme significance of Lovecraft setting so many of his stories in New England' while writing many of them in the first person, he never actually connects Lovecraft's autobiographical geography to the content of his work in any detail (Fisher, 2016, pp. 19–20). This oversight is remarkable because even a cursory consideration would highlight the degree to which Lovecraft's work reflects the same racial and gender ideology as his precursors, but Fisher proceeds without mentioning Lovecraft's well-documented racism, even as he remarks on the degree to which Lovecraft's stories 'are obsessively fixated on the question of the outside: an outside that breaks through in encounters with anomalous entities from the deep past, in altered states of consciousness, in bizarre twists in the structure of time' (Fisher, 2016, p. 16). Fisher's oversight is only compounded by readings of *The Weird and the Eerie* that present this dehistoricizing lacuna as a sign of intellectual independence rather than a straightforward lack of the necessary context:

> One may pause on the political use of the weird (though perhaps not the eerie) that is found in Fisher's text. This is not to say he advocates a politics of the weird, but that there are certain aesthetic and affective themes which cannot but eventually be translated, willfully or not, into the political register. While much ado has been made of Lovecraft's fascism and racism, this has not stopped those who revile his work from banking on his aesthetic-political power. [. . .] Such strategies may perturb the ideological purity of some Leftists, a purity which often results in a paranoia that equates explaining with justifying, engagement with promotion. But these equations are forms of defensive panic which prefer preaching to the choir rather than productively disagreeing.
> (Woodard, 2017, p. 1182)

Such a reading is remarkable because it misses the rather obvious point that addressing Lovecraft's racism makes his work (and thus the weird) more intelligible, not less. This much was made clear by William Hutson of the experimental rap group clipping., who remarked during an interview to promote their 'Afrofuturist, dystopian concept album' *Splendor & Misery* that

> H. P. Lovecraft's cosmic pessimism is only terrifying if you're a straight white man and you thought you were the center of the universe anyway. To anyone else – and this is why his racism comes into it – finding out that you're not the most important thing in the universe is a relief. I think it's interesting that his characters go mad when they figure out that humanity doesn't matter. It's only terrifying if you ever thought you were important, if everything in society has propped you up as the dominant category.
>
> (Burns, 2016)

Hutson's summary will only become more relevant in the pages to follow as the connections between this earlier literary weird and the kept weird of 21st century USAmerican cities becomes clear.

Civic violence

Aside from a brief mention in the first sentence of the 'Foreweird' to *The Weird: A Compendium of Strange and Dark Stories* (Michael Moorcock in VanderMeer and VanderMeer, 2012, p. 1), weird fiction in the United States has not been considered alongside the 'Keep [City] Weird' slogan, even when the latter has been examined in an academic context (Long, 2010). In fact, there are meaningful interactions between the two, and the kept weird of civic slogans should be considered a particular iteration of the wider phenomenon known as 'the weird'. That said, it will be most instructive to begin not with the kept weird itself, but rather the political violence that helps define its contours, as seen in three of the various cities closely associated with the weird.

On March 20, 2018, one day before Mark Anthony Conditt killed himself in the suicide bombing outside Austin, Texas that ended his nineteen-day spree, the Associated Press published an article asking, 'Can Austin stay weird despite the bombs that keep exploding?' (Weissert and Vertuno, 2018). Published even before the bomber's identity was known and his association with an extremist evangelical Christian organization made clear (Nashrulla and Jamieson, 2018), the article suggested that the weird of Austin might be the very reason for the bombs in the first place:

> The blasts have sent a deep chill through a hipster city known for warm weather, live music, barbeque and, above all, not taking itself too seriously. Could all that make Austin, whose population and economy are booming, whose politics are liberal and whose diversity is rich more likely to be targeted?
>
> (Weissert and Vertuno, 2018)

Before he died, Conditt, who was not from Austin, but rather the neighboring town of Pflugerville, recorded a 25-minute video discussing the bombings, but police declined to release the recording and have only given summary statements regarding its contents, with the Austin Police Chief Brian Manley describing the video as 'the outcry of a very challenged young man talking about challenges in his personal life that led him to this point' (Fernandez et al., 2018). Nevertheless, Conditt's time in an evangelical home-schooling program and the survivalist group Righteous Invasion of Truth (RIOT), as well as his political writings opposing women's bodily autonomy and homosexuality (Nasshrulla and Jamieson, 2018) suggest he was quite clearly aligned with precisely that same reactionary white male ontology that characterizes an aversion to the weird in fiction.

In 2013, a similar connection between violence and the civic weird appeared in an article about Santa Cruz, California, asking if the 'offbeat branding effort [has] gone too far?' following the deaths of two police officers, the first in the city's history (Hoppin, 2013; Pasko and Brown, 2013). While the article suggests that the deaths could be tied to efforts to 'Keep Santa Cruz Weird', floating the notion that 'the city has gone too far nurturing its offbeat reputation, trading order for chaos, and becoming an asylum for the troubled and the wicked', the identity of the killer belies this easy connection between the weird and violence (Hoppin, 2013). The killer was Jeremy Goulet, a former United States Marine and US Army helicopter pilot that prior to his double murder had been court-martialed for multiple counts of rape and arrested multiple times for privacy invasion, attempted murder, assault, and drunk and disorderly conduct (Santa Cruz Sentinel, 2013). Far from being a member of the 'freaks, hippies, surf rats, pothead programmers, environmental hardliners, lefties, cultists, druggies, punks and dropouts' associated with the Santa Cruz weird (Hoppin, 2013), Goulet was a stereotypical picture of white male violence, not an aberration but the norm for the imperial white supremacist capitalist patriarchy of the United States.

As the epicenter of multiple white supremacist groups, recent eruptions of violence in Portland, Oregon have at least led to slightly more nuanced accounts of the relationship of violence to the city's particular version of the kept weird. Specifically, Portland-based expert on white supremacy Randy Blazak has remarked,

> The nature of Portland is we foster those at the margins. This is 'Keep Portland Weird' political science version. We like the people who aren't identified with mainstream business as usual life, whether that's in music or fashion or politics. We have a lot of anarchists in the city. We also have a healthy dose of extremist libertarians. There is this celebration of the people at the margins. Sometimes, we get the people we don't like at the fringe, as well as those we do like.
>
> (Parks, 2017)

However, even Blazak's more nuanced suggestion that a 'celebration of the people at the margins' might allow space for reactionary white male violence to fester is belied somewhat by the degree to which this violence is not at the fringe, but rather (as always) represents the violent maintenance of a particular notion of white USAmerican identity. For example, the far-right group Patriot Prayer is led by Joey Gibson, who as of October 2018 is the Republican candidate for a state senate seat in neighboring Washington (The Grouch, 2018). Gibson seems to have explicitly targeted Portland precisely because of the degree to which it appears weird, writing in advance of a planned rally,

> [T]he stench-covered and liberal-occupied streets of Portland will be CLEANSED. CLEANSED, I say.
> The streets will be flowing with freedom, and the air will be filled with patriotism. And fear will have no place in our midst . . . Recourse will be swift, for those who wish to oppress our freedoms . . . And the hands of Justice shall smite them with a vengeance heretofore unknown to these ne'er-do-wells. Join me, Patriots . . . let the memories and struggles of our Founding Fathers not be in vain.
>
> (The Grouch, 2018)

That this apocalyptic rhetoric might be tied to actual violence is made obvious by the case of Jeremy Christian, a fellow Patriot Prayer member who murdered two people after they tried to stop him from assaulting two teenage girls on a Portland train (Parks, 2017; The Grouch, 2018). As Kazak makes clear, the murders committed by Christian were, like in the case of Goulet, not an aberration but rather the expected outcome of ongoing processes:

> What is interesting about Jeremy Christian is how many people of color were not surprised. It's a violent manifestation of the things that happen every day in Portland, the gentrification of people of color right out of the city. This is just the latest chapter of Oregon as white man's land. It goes back before the formation of the state. The Oregon Land Donation Act was for white settlers only. There are different manifestations of that, whether it's the Constitution in 1895, or the dominance of the Ku Klux Klan in the 1920s or the skinheads in the 1980s or gentrification in the 2000s.
>
> (Parks, 2017)

What is clear from these stories is that the kept weird of Austin, Santa Cruz and Portland is not simply a case of a particular term's semiotic flexibility, but rather shows that the intersectional conflicts implied by the impulse to keep something weird are the very same that have animated USAmerican weird fiction. What is meaningfully different, then, is the particular ontology centered by these different instances of 'the weird'.

The kept weird

Most obviously, the key difference between the weird of fiction and the weird of civic pride is the moral reversal that takes place as the weird becomes the thing under threat from the outside. However, this movement cannot be seen as a mode of resistance or creative disruption operating in an opposite direction from (for example) Bierce's 'The Eyes of The Panther' when it reverses the relationship between colonizer and the land. Instead, when cities frame 'the weird' as a point of pride, they are simultaneously diminishing the perceived threat posed by those minority ontologies associated with the weird at the same time they commodify these ontologies. That the far-right subsequently targets these cities as emblematic of precisely the fears that motivated the original USAmerican weird is merely the inherent irony of USAmerican liberalism, which by definition can only ever be a liberal version of imperial white supremacist capitalist patriarchy.

That the kept weird is representative of commodification and bourgeois liberalism rather than genuinely 'fringe' ontologies is made painfully obvious by the history of 'Keep Austin Weird'. The slogan was coined by librarian and professor Red Wassenich sometime prior to July 2000, when he decided to donate to a local radio show, explaining his support by saying that the show 'helps keep Austin weird' (Yardley, 2002). He and his wife began selling bumper stickers with the phrase and donating their proceeds to charity (Yardley, 2002). However, almost immediately, the phrase turned from a sign of civic contribution to conspicuous consumption as an apparel company copyrighted the phrase, leading to a series of ongoing legal disputes (Yardley, 2002; Thurmond, 2017). In a post titled '"Keep Austin Weird" – Buy More Stuff", the original coiners reflect on 'the boring irony of the entire "movement"' being co-opted by corporate interests, as the initial deployment of the phrase was 'a small attempt to counter Austin's descent into rampant commercialism and over-development' (Keep Austin Weird, 2018).

Like Austin, Santa Cruz had a reputation for the weird well before the adoption of the slogan 'Keep Santa Cruz Weird' was popularized by Santa Cruz County Board of Supervisors Chairman Neal Coonerty after Austin's kept weird became widely known (Hoppin, 2013). The easiest way to understand this reputation and the various ways it interacts with the weird in general is the 1987 film *The Lost Boys*, because although it premiered a decade before the slogan was first promoted in Santa Cruz, the story of teenaged brothers moving to the fictional town of Santa Carla and battling vampires firmly connected the evolution of weird fiction to the city's longstanding reputation for the weird in general. In particular, the film connects Santa Cruz's reputation for counter-culture and violence – including a four-year period that saw multiple serial killers operating in the area (Dowd, 2018) – to comic books, an important feature of weird fiction generally overlooked in extant discussions of the genre.

The Weird and the Eerie makes mention of comic books only once, when it notes during a discussion of how 'Lovecraft's stories are full of thresholds between worlds . . . Gateways and portals routinely feature in the deeply Lovecraftian stories of the Marvel Comics character Doctor Strange' (Fisher, 2016, p. 28). Similarly, the introductory essay to a 2017 issue of *Textual Practice* concerned with weird fiction mentions comic books only once, when it claims that following the death of H.P. Lovecraft in 1937,

> the weird then sank lower into the twilight territory of horror comics, with titles like *Weird Chills*, *Weird Horrors*, *Weird Science*, *Weird Tales of the Future*, a boom that began in 1949 and was abruptly curtailed by a moral panic and self-censorship in 1954,

suggesting that comic books' relationship to the weird can be summed up in those five years (Luckhurst, 2017, p. 1044). In both the 'Foreweird' and Introduction to *The Weird: A Compendium of Strange and Dark Stories*, a 'stun-your-enemies huge' collection 'covering over a century's worth of fiction', comic books are not mentioned at all (VanderMeer and VanderMeer, 2012, pp. 1–15; Carroll, 2015). One possible reason for this lingering oversight is an overemphasis on the word 'weird', which does not account for comic books' often audacious relationship with copyright law as it relates to the use of synonyms. *The Weird and the Eerie* almost touches on this lexical shift when it cites Doctor Strange, who first appeared in the comic book *Strange Tales*, a title that began publication in 1951 and went on to feature some of the most important comic book writers and artists of the 20th century.

The Lost Boys repeats this synonymous link to the weird in one of its early scenes, as brothers Michael and Sam arrive in Santa Carla for the first time as a cover of The Doors' 'People Are Strange' plays. From there the connection to weird fiction only grows, as Sam befriends the Frog Brothers, two other teenagers whose parents own the local comic book store. Scenes at the comic shop were filmed at Santa Cruz's real-life Atlantis Fantasyworld before the location was destroyed by an earthquake in 1989, though the store has since relocated (Atlantis fantasyworld, 2018). Even before the slogan 'Keep Santa Cruz Weird' was adopted, the film simultaneously served to tie the city to the weird in general while priming this association to become a site of commodification and gentrification.

That Santa Cruz's weird has been fully co-opted by corporate interests is made clear by Kevin Cornell, the vice president of a software company who responded to the question of whether Santa Cruz's weird had gone 'too far' by remarking that 'unfortunately, it's attracted the dangerous part of weird as well as the good stuff' (Hoppin, 2013). For Cornell 'the dangerous part of the weird' means 'beaches closed due to needles' and 'a homeless camper' living on his property, but he imagines that the city can still benefit from 'the good stuff' so long as it enforces semiotic boundaries:

> I don't think weird is the problem if we define it as what we want it to be. It's the definition and the rules you wrap around it . . . We don't have any containment (now), we just run around saying we're weird.
>
> (Hoppin, 2013)

That what was once considered outside the bounds of understanding or even language might now be defined by 'the rules you wrap around it' emphasizes the degree to which the kept weird represents the capitalist co-option of what might have previously been considered radical or threatening.

As Blazak (Parks, 2017) indicates in his analysis of Jeremy Christian's Portland murders, Portland's weird and the violent reprisals against it are inextricable from an ongoing process of gentrification that goes back at least two decades, roughly to when the city adopted its own version of the kept weird slogan. Taken together with the cases of Austin and Santa Cruz, it seems the adoption of the 'Keep [City]

Weird' slogan performs precisely the opposite action the phrase implies, as the slogan appears precisely at those points when a city's unique character has begun the process of smoothing away any features that might upset the sensitivities of bourgeois liberalism or the friendly capitalism it embraces. Thus, the kept weird is truly kept in the sense of a transactional relationship, wherein ostensibly fringe identities are allowed within the bounds of civic culture so long as they can be made safe and sanitized for consumption by the willing subjects of imperial white supremacist capitalist patriarchy.

References

Atlantisfantasyworld.com (2018) 'Atlantis Fantasyworld'. Available at: www.atlantisfantasyworld.com/lostboys.htm [Accessed 3 Oct. 2018].

Bierce, A. (1897) 'The Eyes of the Panther', *The San Francisco Examiner*.

Bierce, A. (2018) 'An Inhabitant of Carcosa | Weird Fiction Review', *Weird Fiction Review*. Available at: http://weirdfictionreview.com/2014/03/an-inhabitant-of-carcosa/ [Accessed 3 Oct. 2018].

Burns, T. (2016) 'Clipping. on Afrofuturism, Fast Rapping and Sound Design', *RBMA Daily*. Available at: http://daily.redbullmusicacademy.com/2016/09/clipping-interview [Accessed 2 Oct. 2018].

Carroll, T. (2015) 'Weird Fiction: A Primer', *Literary Hub*. Available at: https://lithub.com/weird-fiction-a-primer/ [Accessed 3 Oct. 2018].

Dowd, K. (2018) '"Murder Capital of the World": The Terrifying Years When Multiple Serial Killers Stalked Santa Cruz', *SFGate*. Available at: www.sfgate.com/bayarea/article/santa-cruz-kemper-mullin-frazier-murders-12841990.php [Accessed 3 Oct. 2018].

Fernandez, M., Saul, S. and Healy, J. (2018) 'Who Is Mark Conditt, the Suspected Austin Serial Bomber?' *The New York Times*. Available at: www.nytimes.com/2018/03/21/us/mark-anthony-conditt-austin-bomber.html [Accessed 2 Oct. 2018].

Fisher, M. (2016) *The Weird and the Eerie*. London: Repeated Books.

George Washington's Mount Vernon (2018) *Conotocarious*. Available at: www.mountvernon.org/library/digitalhistory/digital-encyclopedia/article/conotocarious/ [Accessed 3 Oct. 2018].

The Grouch (2018) 'In Portland, Patriot Prayer & Proud Boys Want Immigrants Heads "Smashed into the Concrete": Gave Nazi Salutes While Screaming Racial Slurs – It's Going Down', *It's Going Down News*. Available at: https://itsgoingdown.org/in-portland-patriot-prayer-gave-nazi-salutes-while-screaming-racial-slurs/ [Accessed 2 Oct. 2018].

Hawthorne, N. (1835) 'Young Goodman Brown', *The New England Magazine*.

Hoppin, J. (2013) '"Keep Santa Cruz Weird": Has Offbeat Branding Effort Gone Too Far?' *The Mercury News*. Available at: www.mercurynews.com/2013/03/16/keep-santa-cruz-weird-has-offbeat-branding-effort-gone-too-far/ [Accessed 2 Oct. 2018].

Keep Austin Weird (2018) 'Keep Austin Weird'. Available at: www.keepaustinweird.com/ [Accessed 3 Oct. 2018].

Long, J. (2010) *Weird City*. Austin: University of Texas Press.

Luckhurst, R. (2017) 'The Weird: A Dis/orientation', *Textual Practice*, 31(6), pp. 1041–1061.

Nashrulla, T. and Jamieson, A. (2018) 'Here's What We Know About the Austin Package Bomber', *Buzzfeed News*. Available at: www.buzzfeednews.com/article/tasneemnashrulla/austin-bombing-suspect-mark-anthony-conditt#.mw2ZLpkag [Accessed 2 Oct. 2018].

Parks, C. (2017) 'Oregon White Supremacist Expert Explains the Changing Landscape of Hate Groups', *The Oregonian*. Available at: www.oregonlive.com/portland/index.ssf/2017/06/oregon_white_supremacist_exper.html [Accessed 2 Oct. 2018].

Pasko, J. and Brown, J. (2013) 'Two Santa Cruz Police Officers, Suspect Shot and Killed', *The Mercury News*. Available at: www.mercurynews.com/2013/02/26/two-santa-cruz-police-officers-suspect-shot-and-killed/ [Accessed 2 Oct. 2018].

Ross, J. (2016) 'Mike Pence on Broad Shoulders, Strength and Leadership', *Washington Post*. Available at: www.washingtonpost.com/news/the-fix/wp/2016/10/06/mike-pence-on-broad-shoulders-strength-and-leadership [Accessed 3 Oct. 2018].

Santa Cruz Sentinel (2013) 'Timeline: The Troubled History of Jeremy Goulet Leading to Shooting of Two Santa Cruz Police Officers'. Available at: www.santacruzsentinel.com/2013/03/01/timeline-the-troubled-history-of-jeremy-goulet-leading-to-shooting-of-two-santa-cruz-police-officers/ [Accessed 2 Oct. 2018].

Thurmond, S. (2017) 'Grudge Match: Keep Austin Weird Trademark Holders vs. Keep Austin Weird Fest Organizers', *Austin Monthly*. Available at: www.austinmonthly.com/AM/November-2017/Grudge-Match-Keep-Austin-Weird-Trademark-Holders-vs-Keep-Austin-Weird-Fest-Organizers/ [Accessed 3 Oct. 2018].

VanderMeer, A. and VanderMeer, J. (2012) *The Weird: A Compendium of Strange and Dark Stories*. New York: Tor Books.

Weissert, W. and Vertuno, J. (2018) 'Austin Bombings Put Chill in City Known for Keeping It Light', *AP News*. Available at: https://web.archive.org/web/20180320220348/https://apnews.com/19bb53fd62cc454ebff9330fa5f225df/Can-Austin-stay-weird-despite-the-bombs-that-keep-exploding [Accessed 2 Oct. 2018].

Woodard, B. G. (2017) 'Book Review: The Weird and the Eerie'. In *Textual Practice*. Vol. 31, No. 6, 1181–1197.

Yardley, J. (2002) 'Austin Journal; A Slogan Battle Keeps Austin Weird', *The New York Times*. Available at: www.nytimes.com/2002/12/08/us/austin-journal-a-slogan-battle-keeps-austin-weird.html [Accessed 2 Oct. 2018].

2 'Something Like a Circus or a Sewer'

The thrill and threat of New York City in American culture

Keith McDonald and Wayne Johnson

> And with the awful realisation that New York was a city after all and not a universe, the whole shining edifice that he had reared in his imagination came crashing to the ground.
>
> – F Scott Fitzgerald, *My Lost City*

New York City exists, as Lou Reed wrote in his song 'Coney Island Baby' (and quoted in the chapter title), as a precarious and often oppositional space, as an illusion and as an all-too jarring reality. It is both a 'geographical entity' and a 'cultural production' (Pomerance, p. 3). New York is also the popular culture city par excellence, with the beguiling otherworldliness of skyscrapers juxtaposed with the dystopic social realities of its street level. Like any other object of popular culture, the city is 'constructed like a text' (Campbell, p. 200); it is an everchanging cultural product that has multiple meanings that raise the question of authenticity and mediated simulation, which in many ways, is where it exists. It is both a living place (culture) and an abstract image (idea). New York City is the location and inspiration for the evocative photography of Alfred Stieglitz; the nation-defining songs of Tin Pan Alley; the improvisation, in music and in lifestyle, of the Jazz age; the political expressionism of the Harlem Renaissance; the kinetic energy in the music of Leonard Bernstein; the creative artistry of the comic book (with Metropolis and Gotham standing in as thinly veiled references to New York); the modern narrative collages of John Dos Passos; the postmodern recycled composites of Robert Rauschenberg; the celebrity art of Andy Warhol's Factory studios; the drag balls of *Paris is Burning* (Dir. Jenny Livingstone, 1990); the raw dexterity of hip-hop and the neon overload of advertising in the Broadway Theatre district of Times Square. New York City is a cultural coliseum, a collision of chaotic and incompatible images and multiple voices that somehow forms a strangely authentic whole.

F. Scott Fitzgerald's view of New York, quoted above, is integral to his masterpiece, *The Great Gatsby* (1925); it is an extremely well-studied and well-worn examination of New York's alchemic charm. Nonetheless, Fitzgerald's ideas are still pertinent to this chapter and to leave him out would be perverse. Similarly, Gatsby as a character is a fiction of his own creation, a reinvention of a damaged yet optimistic and romantic individual. That he exists in Fitzgerald's fictional depiction is also appropriate here, too. He is a simulacra bound within a simulacra which still resonates today, a man caught between two lives, one desperately romantic and yet desperately corrupt, lurching (in a dizzying car and a soothing swimming pool) towards a tragic end. However, he and his peer characters live on in print, theatre productions and films, proving the allure of the power of the narcotic space of New York and its simulated charm, its authentic/inauthentic contradiction. This can be illustrated in Gatsby's library, with its unopened and unread books (and which are in fact hollow cases), a 'heterotopic' space (which we will explore) and a show of style and wealth rather than a base of true depth. This can be seen in the reinterpretation of *The Great Gatsby* in Baz Lurhmann's filmic depiction of the novel in 2013. It is an exercise in postmodern simulation with a disregard for authenticity. This is most clearly accentuated in the use of hip-hop on the soundtrack with the inclusion of the song '100$'. Its incorporation into the film illustrates the filmmaker's take on New York as a manifestation of revolving artifice defined in part by its own matrix of reinvention heavily tethered to commercial gain. This mirrors, in some ways, Frank Sinatra's legendary rendition of 'New York, New York' (in which in the title itself attests to the sumulacra) which ends with the line '*it's up to you* New York, New York . . . New York' (our italics).

In many ways, and like many heterotopic spaces, New York is in a constant state of 'morphication'. Of course, this is the state of many places in world history (the restoration of London emerging out of its great fire and the effect on Bilbao after its radically charged art influx prove this point). However, New York, and in particular Manhattan Island, is arguably the most mediated place on Earth and, as such, is under a magnifying glass as a presence in a constant state of flux, polarised between a historic cultural grounding and a driving force into the future. If considered then as a cultural organism it raises questions about the futuristic and paradoxically elusive nature of the postmodern text, as Brian McHale states:

> 'Postmodernist'? The term does not even make sense. For if 'modern' means 'pertaining to the present', then 'postmodern' can only mean 'pertaining to the future', and in that case what could postmodernist fiction be except fiction that has yet to be written.
>
> (McHale, p. 4)

Jacques Derrida, in *Spectres of Marx* (2014), developed the concept of 'hauntology', where (in this case in relation to the ideological traction of Marx) he formed a notion that cultural thinking is constantly reliant on other previous input and is self-consciously or subconsciously aware of prior incarnations, concepts and correspondences (this is how genres and continual discourses are birthed and maintained). When considering this, the self-conscious nature of New York as a curated yet inconsistent cultural entity, striving towards the new yet melancholically tethered to the past, serves as a useful lens. Derrida also stated:

> That the without-ground of this impossible can nevertheless take place is on the contrary the ruin or the absolute ashes, the threat that must be thought,

and, why not, exorcised yet again. To exorcise not in order to chase away the ghosts, but this time to grant them the right, if it means making them come back alive, as *revenants* who would no longer be *revenants*, but as other arrivants to whom a hospitable memory or promise must offer welcome without certainty, ever, that they present themselves as such.

(Derrida, p. 162)

New York, like many other places, and specifically in terms of Western popular culture, both epitomises the fresh yet nostalgic aesthetic and exists in a state of hauntology, always looking to the past to inform the future in a melancholic state which wants to simultaneously move on, remain static and return to previous romantic states of being. This is perhaps spatially typified by the monument of the Twin Towers in the wake of the attacks of 9/11 named *The Tribute of Light*, which consists of 88 searchlights beamed into the night sky. Fittingly, in the context of this chapter, the monument is both dazzling, beautiful and somehow ethereal. And, considering its form, in search for some future, not unlike the poetic ending of *The Great Gatsby*, which reads,

Gatsby believed in the green light, the orgiastic future that year by year recedes before us. It eluded us then, but that's no matter – tomorrow we will run faster, stretch out our arms farther . . . And one fine morning – So we beat on, boats against the current, borne back ceaselessly into the past.

(Fitzgerald, pp. 171–172)

James Baldwin acknowledged the uniqueness of New York City and pointed out the relative lack of monuments and historic pomposity of other cultural meccas. He did, though, identify a key point relating to the binary tensions which are both fascinating and daunting when he wrote '[A]ll other cities seem at best, a mistake, and at worst, a fraud. No other city is so spitefully incoherent' (Baldwin, p. 7). Although one could of course identify 'traditional' edifices such as the Empire State Building and The Statue of Liberty, we would argue that the steam-filled vents of the city streets, the hot-dog stands and the Staten Island Ferry are equally important to its cultural currency. This is of course relevant to other cities also; London's tube system is iconic and historic, for instance. However, in terms of popular culture, New York seems to have a unique traction with regards to the mix of 'high' and 'low' cultural meshing like no other place in the Western world. This is not always successfully achieved. In a time which has witnessed a new crisis in affordable housing in the city of New York, the recent trend of a 'super-skinny, super-tall' new style of 'skyscraper' is epitomised by 432 Park Avenue. The building was completed in 2015 and stands 425 metres tall with 96 floors, and was built without planning permission or indeed any municipal scrutiny. It is the tallest residential tower in the world at present and embodies a new form of absentee property ownership, standing as empty 'silos of billionaires . . . casting ever-longer shadows across Central Park' which represent a new form of currency in a post-2008 global financial world—the new 'age of technical ingenuity and extreme inequality' (Wainwright, 2019, pp. 34–38).

Ever ironically, these can be compared to Gatsby's empty book 'cases', purporting to be populated.

Michael Foucault explores the notion of what he calls a heterotopia in relation to spaces that are projections of cultural practices, ambitions and, at times, tensions and simulations of authentic experiences. These include spaces such as tropical gardens that simulate other climates, art galleries and other cultural exhibition sites that curate creative expressions and monuments to the past, etc. As perhaps the world's biggest and most certainly most documented space (never sleeping, constantly evolving and yet retained), New York epitomises the heterotopia in the terms which Foucault describes:

> There are also, probably in every culture, in every civilization, real places – places that do exist and that are formed in the very founding of society – which are something like counter-sites, a kind of effectively enacted utopia in which the real sites, all the other real sites that can be found within the culture, are simultaneously represented, contested, and inverted.
>
> (Mirzoeff, p. 231)

Adding to this discussion regarding the cultural imagination in a psycho-geographical sense, Edward Soja posits the concept of a dialogic offspring between material reality and imagined fantasy (which he terms as the 'second-space') which then results in what he calls a 'third-space', which he explains as 'a creative recombination and extension, one that builds on a first-space perspective that is focused on the "real" material world and a second-space (in which) everything comes together in third-space' (Soja, p. 56). Soja further argues that this approach involves acknowledging 'a mode of thinking about space that draws upon the material and mental spaces of the traditional dualism but extends well beyond them in scope, substance and meaning' (Soja, p. 11). This concept is fundamentally a creative one and, as an idea, is impossible to define, as are any city's fundamentally true identities. However, in the case of New York City, this is amplified and broadcast around the globe.

Although these examples represent numerous case studies, which could suitably act individually as demonstrative texts to the ideas presented here, New York is perhaps best understood as a complex hypertext, a mediated bricolage of both contradictory and complimentary texts that have embedded themselves in the cultural imagination. It is a utopian and dystopian polarising entity, best described as an imaginative heterotopian force under constant scrutiny and change (for good or worse). For the writer Thomas Wolfe, New York was

> a cruel city, but it was a lively one; a savage city, yet it had such tenderness; a bitter, harsh and violent catacomb of stone and steel and tunnelled rock, slashed savagely with light, and roaring, fighting a constant ceaseless warfare of men and machinery; and yet it was so sweetly and so delicately pulsed, as full of warmth, of passion, and of love, as it was full of hate.
>
> (Heyes, p. 76)

It is though, through popular culture that this is best captured, as Cohen and Taylor state, these bricolage entities combine in a flow which is enamoured with opportunity:

> All around us-on advertisement hoarding, bookshelves, record covers, television screens these miniature escape fantasies present themselves. This, it seems, is how we are destined to live, as split personalities in which the private life is disturbed by the promise of escape routes to another reality.
> (Cohen and Taylor, in McHale, p. 38)

First, the above accentuates the complex and confusing environments and situations inhabitants must navigate. Second, it pits the urban, multicultural 'human world' against a backdrop of a far more complex and unreadable myth that makes some of the concerns of the tourist petty and fleeting. Third, it makes all viewers, readers, etc. migrants, in that we are all experiencing these languages as non-natives drawn into the mythical 'native' mechanisations.

Clearly, cinema is a particular medium that many have focused on with the goal of producing excessive superfluity in terms of artistic representation. There are an abundance of filmed representations of the city which are incredibly diverse in terms of genre and which have made an indelible mark on film-lore. David Clarke states that cities are particularly cinematic in their qualities because filmmakers can offer multiple interpretations to suit their narrative and mise-en-scene. Clarke further states that the American city is particularly fitted to a hyperreal, cinematic experience and quotes Baudrillard who contends 'to grasp its secret, you should not, then, begin with the city and move inwards towards the screen; you should begin with the screen and move outwards towards the city' (Clarke, p. 1). Depictions of New York in cinema often relish in its dizzying and intoxicating chaos. For instance, in musical film traditions, both *On the Town* (dir. Gene Kelly and Stanley Donen, 1949) and *West Side Story* (dir. Robert Wise, 1961) score and choreograph the narcotic hit of the sensation that New York appears to be. Filmed in glorious technicolour and on location in the city itself, *On the Town* (music by Leonard Bernstein and lyrics by Betty Comden and Adolph Green, although with new songs by Roger Edens), like any number of examples of popular culture, has been through many incarnations; it was initially a ballet by Jerome Robbins called *Fancy Free*, it then became a Broadway musical before becoming a Hollywood musical. It captures the *carpe diem* of three sailors on 24-hour shore leave in the city; New York is presented as an intense high. This idea of constant remaking acts as a suitable metaphor for the place itself, which is an ever-evolving simulation of itself. In addition, there is a moral chaos often depicted in films such as *Working Girl* (dir. Mike Nichols, 1988), *Wall Street* (dir. Oliver Stone, 1987) and *The Wolf of Wall Street* (dir. Martin Scorsese, 2013) which depict a hedonistic mix between big business, play and human consequence, individual status elevation and reinvention. Carl Boggs sums this up with the following statement that embodies 'the New York state of mind:

> The capacity of sprawling business empires, banking systems, governments and international agencies to manage economic, political, and cultural life

coincides, paradoxically, with a civic life that is anything but stable and orderly.

(Boggs, p. 2)

Other genres include horror (*Rosemary's Baby*, dir. Roman Polanski, 1968), the romantic comedy (*When Harry Met Sally*, dir. Rob Reiner, 1987), the sci-fi dystopia (*Escape From New York*, dir. John Carpenter, 1981), the monster movie (*King Kong*, dir. Merian C. Cooper and Ernest B. Schoedsack, 1933), the superhero adventure film (*Avenger's Assemble*, dir. Joss Whedon, 2012; *Spiderman 2*, dir. Sam Raimi, 2002) and the children's fantasy film (*Elf*, dir. John Favreau, 2003).

From *Coney Island* (dir. Ralph Ince), a silent comedy made in 1928, through to the recent *Ready Player One* (dir. Steven Spielberg, 2018), New York is represented as a location of chaos. In *Ready Player One*, set in 2045, the world can escape the dystopic realties of the present by retreating into their imagination, represented by OASIS, a virtual universe that provides freedom for its players. On the hunt for an 'Easter-egg' as part of a treasure hunt, Wade Watts engages in a race through New York into Central Park, accompanied by the mayhem inflicted by King Kong and Godzilla. In the film, the whole of New York City becomes a virtual-reality city—a roller coaster ride. This chaos is, of course, both alluring and threatening, which echoes the iconic scenes in the original *King Kong* where the intoxicating spectacle (at street level) of the unveiling of the exotic creature all too quickly leads to violence, threat and fatality in a dizzying celluloid fantasy. Kong himself is anything but ordinary, and it is no surprise that the thrilling reveal and the calamity that follows take place atop the Empire State building, whilst others look on speechless as Kong roars amid this celluloid spectacle. As Foucault states:

> Heterotopias are disturbing, probably because they secretly undermine language, because they make it impossible to name this and that, because they shatter or tangle common names, because they destroy syntax in advance, and not only the syntax with which we construct sentences but also that less apparent syntax which causes words and things (next to and also opposite one another) to 'hold together'. This is why utopias permit fable and discourse: they run with the very grain of language and are part of the fundamental dimension of the fabula; heterotopias . . . dessicate speech, stop words in their tracks, contest the very possibility of grammar at its source; they dissolve our myths and sterilize the lyricism of our sentences.

(Foucault, pp. xvii–xix)

This common lexicon, is, of course, reliable yet ultimately dull, which so many depictions of the city rally against for romantic, exhilarating and sinister outcomes. *On the Town* symbolises freedom for the three sailors; in stark contrast, *Taxi Driver*'s (dir; Martin Scorsese, 1976) New York is a hellish prison for Travis Bickle (and literally so in the case of John Carpenter's *Escape from New York*). In *Taxi Driver*, Bickle, a damaged Vietnam veteran, drives a taxi at night through the decaying, dystopic streets of a New York of nightmares, encountering other misfits, the seedy and the corrupt. The film explores Bickles' insomnia-fuelled alienation from society and his growing sociopathic obsessions as he attempts to assassinate a liberal presidential candidate. Failing in that act, he becomes, bizarrely, a vigilante hero for killing undesirables—yet again demonstrating the polarised tensions of the city itself.

Much has been made of the 'urban renewal improvements' made by Mayor Ed Koch in the 1980s (he was mayor from 1978–89). Whether it was more his imaginative rhetoric (his 'how'm I doing' slogan) rather than his actual policies which accounted for the transformation in public confidence in the city is difficult to gauge, as in reality, the homicide rates remained high throughout his time. Indeed, it was Rudolph Giuliani, the Republican Mayor from 1993 with his 'broken windows' crackdown on anti-social behaviour which led to a greater social transformation and addressed the root causes of crime; in the process it reduced the murder rate in the city by nearly 75%. The 'I love New York' logo, also designed in the 1970s, although to promote tourism through the state rather than the city specifically, showed how a brand could create a popular emotional response and a self-definition and re-appropriation by inhabitants of Manhattan itself.

Considering the dystopian depiction of New York in the 1970s and early 1980s (which Koch claimed to be solving) in films such as *Serpico* (dir. Sidney Lumet, 1973) and *Fort Apache: The Bronx* (dir. Daniel Petrie, 1981), it would be safe to assume that New York in the pre-Koch era (and in many ways during it) is a place to be endured, wary of and ultimately escaped (as referenced in John Carpenter's take on the matter). However, revealing its inherently contrary nature(s), New York is often depicted culturally across media as a magnetic place to be sought out as an environment for progression and transgression. This is no more cinematically evident than in the work of filmmaker Mike Nichols, who clearly saw New York as a place of transfiguration in terms of post-war developments with regards to the politics of gender and sexuality. This commitment, in a vast body of work which includes a mass of theatre, can also be exemplified in three films: *Carnal Knowledge* (1971), *Working Girl* (1989) and *Angels in America* (2003). In these films, New York is the theatre on which a discussion about sexual politics takes place; a prism by which social change, struggle and negotiation is performed on screen. What is key here and in much of Nichol's other New York-based material (perhaps most clearly seen in *Working Girl*) are the ways in which identity politics are so closely connected to commerce, as seen in the huge success of the TV show *Mad Men* (2007–2015) which could not have been as successfully rendered outside of New York and whose central character is an attractive fraud. In this context, New York is as much of an imaginative destination as it is a geographical one. This is psycho-geography through art, sound, vision and storytelling, albeit a highly monetized one.

What can be further observed, moreover, in American popular culture set in New York, ranging from advertising to any number of Hollywood movies, is an ever-shifting perspective. There is the abstract built environment of the city, as seen from above, where 'the skyline becomes a "storied" place' (Sanders, 115) providing a façade of moral certainties, while the 'awkward and unromantic realities' of the street level are, more often than not, portrayed as a war zone with its festering social ills (Campbell, 204). What also emerges in these films is the juxtaposition of the skyscraper, as 'the architecture of corporate capitalism' (Lindner, 19), and the 'public' space of the street level; the former, as in *Avengers Assemble* and *Spiderman 2*, 'disconnecting people from traditional ways of life' (Sanders, 115). As such, it is the thrill and the threat of New York that has been reflected in Hollywood's representations of the city. In this sense, semiotics and ocular tourism are key in a postmodern sense in that simulation is at their core. Most of us don't physically visit New York, yet culturally and visually we have visited the place even though in an inauthentic manner. As John Urry states, 'generally, we are well aware that most tourism involves, at least in part, the activity of *sightseeing*. In most discourses surrounding travel, there is an emphasis on the centrality of the seeing and collection of *sights*' (Urry, p. 176). This is no more amplified to the level of the hyperreal as it is in cinema's infatuation with New York.

The French high wire artist Philippe Petit recorded his first impressions of New York in the early 1970s in his diary: 'it's old, it's dirty, it's full of skyscrapers, I love it' (Petit, 12). Petit had arrived into a squalid New York of fiscal stagnation (which would culminate in the New York blackout in 1977), abandoned buildings, high crimes rates, drug problems, gang warfare, police corruption and the institutionalised prejudice epitomised in *Serpico* (1973), starring Al Pacino, and the landmark TV series *Kojak* (1973–8). The Twin Towers of the World Trade Center seemed to represent a physical denial of this decline; they were 'vertical constellations' (Lindner, p. 21). In his memoir *To Reach the Clouds* (2002), Petit wrote about how he planned and then executed the audacious, and illegal, tightrope walk 1350 feet above the ground between the Twin Towers in 1974 (the building of which had been completed the previous year at a cost of $1.5 billion). The so-called 'artistic crime of the century' was later made into the documentary film *Man on Wire* (dir. James Marsh, 2008) and the 3-D feature film *The Walk* (dir. Robert Zemeckis, 2015); fittingly, a cultural reinvention from memoir to documentary, through to feature film. Petit spent nearly an hour on the wire, walking back and forth, kneeling and lying down. What he performed would have remained nothing more than a spectacular circus act were it not for the tragic events of September 11, 2001 and the terrorist attacks on the Twin Towers when the towers collapsed, with over 2000 casualties. In his book on the visual representations of 9/11, Thomas Stubblefield argued that 'the paradox of the visual culture of 9/11', in which media coverage Hollywood-ized and spectacularised the event itself 'foregrounds the image and the visual experience in general', but also, and because of the 'unrepresentable' nature of the disaster, 'steeps the events of that day in absence, erasure and invisibility' (Stubblefield, pp. 5–7). Thus, 'le coup', as Petit termed his wire walk of 1974, has now assumed celestial significance. Petit had read about the plans to build the Twin Towers in a French newspaper in the 1960s. From the beginning, his dream was an illusion. As he stated in the documentary, 'the object of my dream doesn't exist yet.' The walk itself was inspired by Petit's refusal to acknowledge that New York could only be seen from the street level, since the Towers, as they certainly appeared to him for the first time he saw them in a *Paris Match* magazine article in 1972, were 'already out of reach' (Petit, p. 8). And, on physically seeing the Towers for the first time, Petit perfectly described the tension as he stared up, between 'the vertical aluminium panel' of the buildings' facade climbing up 'into azure' (elevating, thrilling, clean, aspirational) and 'the horizontal concreter slab' of the ground he stood upon (mundane, threatening, dirty, desolate). He described the tower as 'a landing field for extra-terrestrial vessels' or a 'limitless runway into heaven'. Either way, he went on, 'it is definitely not man-made, nor of any use to us humans'; and yet humans engineered their destruction (Petit, p. 14). As one of his co-helpers on the roof stated in the documentary, 'I remember the vastness of New York. It was magnificent. And the sounds as well, the police sirens all night long . . . It was all so alive. We were kings!' And, on reaching the rooftop, the city, for Petit, simply vanished, and humanity, with all its social, economic and political ills, had 'ceased to exist', as Petit was gracefully balanced between earth and heaven, his material tightrope—a wire cable—serving as an exquisite metaphor for the equilibrium (Petit, p. 15).

In important ways, then, the verticality of New York allows it to be a 'playground for superheroes', or, as in the case of Petit, a super human, or a coliseum for monsters and villains, and an all too real hell for its inhabitants (Sanders, p. 120). In *Avengers Assemble*, for example, the action is played out in the abstract, high above street level, as if the superheroes are on fairground rides. Action is unconnected to street life; indeed, the mayhem is 'scarcely visible from the street' as civilians are evacuated (Sanders, 115). Upper buildings offer 'a whole landscape of ornament, statuary . . . temples, obelisks, pyramids' (Sander, p. 115), but 'without any exposure to its interior reality' (Sanders, p. 97). In *Cloverfield* (dir. Matt Reeves, 2008), New York is well and truly 'under siege'. Presented as previously lost film footage held by the U.S. Department of Defense and deliberately mimicking the horror of 9/11, unknown and half-glimpsed monsters wreak havoc on Manhattan, bringing floating debris, panic, bewilderment and then evacuation to the local populace. Again, there is the curious juxtaposition of the differing sensation of the street level and the subway underground with the roofs of buildings as Rob, the lead character, and his friends search for Beth, his ex. It is as if, and according to Sanders, 'New York's skyscrapers seemed designed not for human beings . . . but for some sort of new race', echoing Petit's reflection (Sanders, p. 100). Clearly the filmmakers wished to replicate the sensation of theme park rollercoaster rides—thrilling as well as frightening (but with an added sense of horror)—since there is no coincidence in the fact that the discovered camcorder film is framed between previous footage taken by Rob of him and Beth talking about Coney Island, the leisure and beach destination on Long Island, then on the way to it, then 'actual' footage of the couple at said island, where Beth states finally: 'I've had a good day'. A simulacrum over a simulacrum.

Film theorist Murray Pomerance stated that 'New York is an eidolon, surely-an image which is possessed of a phantasmatic, apparitional, haunting quality, and that rests out of history as a mark of aspiration, memory and direct experience'

(Pomerance, p. 4). Cinematically and in many other incarnations (or perhaps incantations), New York in representations provides us with the artistic display of a polarised dream mixed with a nightmare vision of hyper, urban America. The haunting quality here is worth attention. Derrida maintains that the political figure looms large on most political discourse after the publication of *The Communist Manifesto*. This concept has of course been enveloped in the wider cultural studies field in terms of discursive patterns. To extend the metaphor of hauntology, if Marx exists as a figure of looming gravitas in a gothic scenario, depictions of New York enter a melting pot of spectral voices and images (perhaps stirred on occasion by Fitzgerald), each pulling in their own disparate directions but ultimately contributing to the perverse disparate whole.

The final example of the heterotopic realisation of New York is the extraordinary performance by Alicia Keys featuring Jay-Z of her song "Empire State of Mind" delivered in Times Square on October 9, 2016. There has never been a more demonstrative live example of space as 'text' other than in fiction and in particular science fiction. The stage on which they performed was simply a part of a wider canvas where buildings, vast screens and other platforms converged in a cacophony which demonstrated the place as both a dizzying chaos (utterly manufactured) and somehow a coherent whole as part of a media explosion (bizarrely authentic). The whole event was a finely crafted escapist dream in terms of the lyrics, music, production and distribution in the digital form of a virus. Typically, footage involves an immense amount of product placement of the most American of recognised products (Coca-Cola, Budweiser, etc.). Of course, after being a continuation of the ongoing cultural voice (as 'heteroglossic' as a voice can be) which includes Literature, Visual Arts, Music and Film and TV, the performance found a huge audience online with over 13 million views to date. What is interesting and poignant is the amount of the audience seen recording their experience on smartphones, creating a simulation of a simulation, which in itself is a serpentine space that defies and defines the postmodern mediated environment. New York is perhaps best thought of as a self-generating and multifaceted fiction, an organism that at least culturally is in a constant state of metamorphosis. McHale makes a point regarding postmodernism and narrative, but this is equally applicable to this city as a text. He writes,

> Typically, in realist and modernist [fiction, the] spatial construct, is organised around a perceiving subject, either a character or the viewing position. The heterotopian zone of postmodernism [narratives] cannot be organized in this way, however. Space here is less constructed than deconstructed by the text, or rather constructed and deconstructed at the same time.
>
> (McHale. p, 45)

Perhaps New York is best considered as a fiction in this sense, existing in resonating ways in cultural imaginations, elusive yet memorable and destined to respawn whilst retaining a sense of history and a glamour and squalor that is now encoded into its DNA. Contradiction is the life-breath of New York in artistic, iconic and iconoclastic senses (being both authentically American and simultaneously utterly fake). It is fixed yet fluid, past yet futuristic, and, fundamentally, un-fundamental.

Works cited

Baldwin, J. (1994) *Just Above My Head*. Penguin.
Boggs, C. (2003) *A World in Chaos: Social Crisis and the Rise of Postmodern Cinema*.
Baudrillard, J. (2005) *America*. London: Verso, 1988, in Clarke, D. (ed.) *The Cinematic City*. Routledge.
Campbell, N. and Kean, A. (2012) *American Cultural Studies*. Routledge.
Derrida, J. (1998) *The Derrida Reader: Writing Performances*. University of Nebraska Press.
Foucault, M. (1970) 'Les mots et les choses: Une archéologie des sciences humaines', Gallimard, 1966; translated *The Order of Things: An Archaeology of the Human Sciences*. Pantheon Books.
Fitzgerald, F. S. (1926) *The Great Gatsby*. Penguin.
Fitzgerald, F. S. (1965) 'My Lost City', in *The Crack-Up with Other Pieces and Stories*. Penguin.
Lindner, C. (2015) *Imagining New York City: Literature, Urbanism, and the Visual Arts, 1890–1940*. Oxford University Press.
McHale, B. (1989) *Postmodernist Fiction*. Routledge.
Mirzoeff, N. (ed.) (2002) *The Visual Culture Reader*. Psychology Press.
Pomerance, M. (ed.) (2007) *City That Never Sleeps: New York and the Filmic Imagination*. Rutgers University Press.
Reed, L. (1990) *Coney Island Baby, from Walk on the Wild Side: The Best of Lou Reed*. RCA.
Sanders, J. (2003) *Celluloid Skyline: New York at the Movies*. Knopf.
Soja, E. (1996) *Thirdspace: Journeys to Los Angeles and Other Real-and-Imagined Places*. Blackwell.
Stubblefield, T. (2015) *9/11 and the Visual Culture of Disaster*. Indiana University Press.
Urry, J. (1992) 'The Tourist Gaze "Revisited"', *American Behavioral Scientist*, Nov.–Dec., 36(2).
Wainwright, O. (2019) *Guardian Weekly*, 13 Feb.
Wolfe, T. (2007) 'The Web and the Rock', in Hayes, W. (ed.) *City in Time: New York*. Sterling.

3 "That Chinese guy is where you go if you want egg foo yung"

Construction and subversion of exotic culinary authenticity in David Wong Louie's *The Barbarians are Coming*

Jiachen Zhang

Introduction

During the period between the 1950s and the 1970s, there was a widespread U.S. fascination with Asian and Oriental cookery among white American housewives. As Mark Padoongpatt (2013) observes, the publication of the first few Asian cookbooks in America facilitated the cultural phenomenon. Rather than being written by Asians or Asian immigrants in America, however, almost all of them were written by American white women who travelled across the Pacific as military wives. In reading these cookbooks, Padoongpatt asserts that this jointly discursive, selective and often prejudiced constitution of culinary authenticity around Asian food ensured the adaptation of Oriental food cultures to the ideological appetites of white Americans during the Cold War, the era that reconfigures Asian Americans from the yellow peril into model minorities who were regarded to be ethnically assimilable. By authenticating their versions of Asian food practices, the American women authors fetishised the culinary differences delivered in Asian cuisines and, more importantly, created their romantic version of "authentic" Asian identity to reiterate their role in U.S. global expansion in Asia. This cultural phenomenon leaves large spaces for us to consider the specific ways in which the U.S. pursuit of authentic Asian taste works to enact various mechanisms of hegemonic politics that result in the material consequence of racialisation and circumscription of Asian subjects. Moreover, are these subjects as invisible and powerless as they seem when overwhelmingly marginalised by U.S. authority over defining Asian Americans and their foodways? How do they respond to the circularity of authenticity and exoticism that expects to limit them to assigned ethnic characters?

To explore the complex power dynamics within this cultural discourse, this chapter discusses the representations of culinary authenticity in Chinese American writer David Wong Louie's critically unknown novel *The Barbarians are Coming* (2000). Set in 1979, the rites-of-passage narration revolves around Chinese American second-generation immigrant Sterling Lung, a former resident chef for the Richfield Ladies' Club and later a host for a cooking show. The novel illustrates how the American patrons establish the essentialist association between the hero's Chinese identity and his limited capacity to cook only authentic "Chinese" food. While Louie delineates the racial and sexual repression of Sterling's subjectivity, adding more complexity to the Cold War enchantment in Oriental food, the novel is deeply concerned with the liberating potential of the culinary authenticity that works to empower the protagonist's impeded identity. Sterling's performance on the cooking show confirms the affective power of exotic culinary authenticity to enact mediation and resistance. Louie focuses on how cooking "authentic" food gives rise to a collective cultural memory, contributing to the formation of a renewed Asian American identity. Louie's interrogation of culinary authenticity, I argue, defies the simplified reading of the cultural phenomenon, frames culinary performances as traceable embodied practices and reveals the need to re-examine the mimetic linkage between culinary authenticity and exotic othering.

My focus on the tension between culinary authenticity and its definition of ethnic origin in *The Barbarians are Coming* builds upon a series of critical works that aim to deconstruct the material and emotional meanings produced in the ethnic encounters. My analysis is informed by Lisa Heldke's (2003) discussion of Euro-American fascination with authentic exotic cuisines. In *Exotic Appetites*, Heldke (2003, pp. 23–44) suggests that the desire thrives on ideas of the essence or the purity of origins that are conventionally understood to be static and unchanging. In my reading, *The Barbarians are Coming* illustrates the ways in which the ability of the U.S. to define the authenticity of an ethnic cuisine allows the food adventurers to not only legitimise their own knowledge, enhance the "cultural capital" under Pierre Bourdieu's (1984) term, but also to access the strangeness of the foreign culture. While U.S. hegemonic citizenship is arguably built upon the quest to authentically connect with often left-behind countries and leaves the weak parties displaced and exploited, Andrew Warnes' (2013) exploration of the historic invention of authentic U.S. Southern barbecue renews our understanding of the complex dynamics of invented tradition and inventions. Culinary authenticity might grow in the hands of powerless culture, leading to a possible threat against a Western ideal of purity. Warnes' emphasis on the affective power of authenticity leads to my contention that in Sterling's cooking and eating, "authentic" food constitutes a disruptive lens to mark his subversive strategy of culinary performances and thus challenge American purist politics of the cultural discourse.

With all this in mind, the chapter investigates the workings of U.S. fascination with exotic culinary authenticity in asserting consuming citizenship, and demonstrates the varieties of Chinese American subjectivities in subverting the cultural discourse from within. I will start to examine the ways in which the American patrons' quest for "real" Chinese food prepared by Sterling legitimises the U.S. hegemonic construction of exotic appetites, metaphorically consuming the Chinese subject by the tropes of food tourism and racial gendering in 1970s America. Yet this circumscription of the hero is soon curtailed by the liberating potentials of perceived authentic Chinese cuisines. This chapter suggests that Louie's portrayal of "egg foo yung" infuses subversive power in the ethnic subjectivity of the hero in a culinary term, owing to the food's strategic initiation of invented tradition to satisfy U.S. exotic appetites. I conclude with an analysis of the connection between culinary authenticity and collective cultural memories in the novel, arguing that the scenes of recognizing once-rejected Chinese gustatory desire evoke the recuperation of collective memories of exploited Asian labourers and facilitate the hero's integration into the Chinese diasporic community.

U.S. construction of exotic appetites

Inspired by a poem with the same title written by Chinese American poet Marilyn Chin, *The Barbarians are Coming* demonstrates the process in which U.S. dominant culture designates Chinese food and culture as exotic. Chin's poem mimics prejudiced views of the Chinese that have existed in U.S. culture, and likens the Chinese race to animals like "horse" and "bison," which are regarded as "equally guilty" (Louie, 2000, p. 1) since they are excluded from the Western civilisation. Louie highlights such a designation of the Chinese as perpetually foreign to the U.S. semiotic order so the American patrons can claim culinary authority over Sterling's cooking. In serving a fundraising dinner held for Mr. Drake's political campaign, the chef learns that the president of the club, a housewife named Libby Drake, and several other women, take no interest in his culinary skill of French cooking but in the consumption of the "authentic" food and his Asian body. Towards the end of the dinner, Libby introduces Sterling as "our very own Chinese chef" (Louie, 2000, p. 47). More intriguingly, she draws the diners' attention to the roasted swans in the main course by labelling them as a "real" Chinese dish without Sterling's consent, even though the swans are prepared in a French style (Louie, 2000, p. 38). As their fascination with Chinese food and chefs illustrates, U.S. desire for cultural difference is politicised through aesthetic value that can often be measured in terms of the exotic. The enticing allure of exotic dishes can not only enrich American palates but also, as suggested in Graham Huggan's (2001, p. 17) observation of exoticism in the context of postcolonial literature, "relieve its practitioners . . . from the burdensome task of actually leaning about the 'other' cultures." The very essence of the fetishised representations of an exotic "other" points to the demystification of foreign cultural forms based on sanctioned and artificial knowledge that grants the dominant culture imagined access to the cultural others.

Mr. Drake's subsequent discussion of his recent travel to China after eating "authentic" Chinese food soon confirms the value of the exotic food, which enables him to increase his social value by performing the newly bestowed "expertise" before others in the West. It is notably infused with a set of discriminating perceptions and sentiments of culinary tourism and adventures. He tells the diners that U.S. evangelical politics applies to the "Ping-pong" diplomacy in which, as revealed by the politician, the American sportsmen intentionally yield to the Chinese counterparts in order to gain their affection. Yet he also holds conflicting viewpoints that Ping-pong is "suited for the whole race of [Chinese]" because "those petite paddles and little balls are perfect for their little hands." Drake attributes such diminutive Chinese figures to the country's collective culture, which, as he claims, removes all "evolutionary imperative" for them "to develop bigger, stronger bodies" (Louie, 2000, p. 50). This caricature of social Darwinism takes for granted the exclusive superiority of U.S. sportsmanship and culture. It also prefigures a mode of contemporary tourism that approaches Asian countries by selecting, consuming and even creating "authentic" Eastern resources and experiences to nourish the mainstream Orientalist expectations. Or, in the butcher and then waiter in Fuchs's metaphorical reflection, the American diners are a group of "Pilgrims" (ibid, p. 41) on the *Mayflower* led by Captain Drake, whose imagined navigation across the Asian continents replaces the unknown Chinese cuisine and culture with his sanctioned and selected perceptions that are necessary to explore the uncivilised land.

Furthermore, Louie expands this Orientalisation of culinary authenticity by delineating the American women's racial gendering of the chef in their club activities, which are expected to compensate for their inflicted gender weakness. Libby Drake, like many other upper-class housewives in the club, is deeply unsatisfied with assigned gender roles in domestic families and public society. Her blunt confession that "I feel so small tonight, with men in my club" (ibid, p. 39), reinforces U.S. historical limitations of women to domestic roles and their subsequent silencing in public affairs during the Cold War. While this comment confirms Sterling's rejection from American normative masculinity in the housewives' imagination, the ladies' feminisation of the Asian chef further interlinks the subjugation of Chinese labourers with the sexual dominance of their bodies. In an evocative scene when Libby accidentally loses her balance during Sterling's preparation of roasted swans, the American housewife "touches [his] shoulder, then caresses [his] ponytail, her fingers running through my hair like a litter of nesting mice" (ibid, p. 40). Whereas the protagonist had hoped his hairstyle would express his masculine virility, relating him to other forebears like the Beatles or Jerry Garcia as a gesture of inclusion into American hippie culture, it becomes a significant site where ethnicity, gender and class interlock to refuse such efforts and associate him with the exotic, ornamental and animalistic. His girlfriend, Bliss, regards his hairstyle as a way of "honour[ing] his forebears" since "this is the way Chinese men have traditionally worn their hair" (ibid, p. 40). What Bliss craves is in fact called a "pigtail," a compulsory hairstyle for Chinese males during the Qing Dynasty. Yet her craving for this "authentic" way of wearing hair immediately invokes the long-standing myths and ideologies that associate Chinese people with certain stereotypical perceptions of being old-fashioned and pedantic. As suggested in Rachel K. Bright's (2017, p. 561) recent exploration into the cultural history of Chinese hair, the word "pigtail" was widely spread in English during the nineteenth century and became highly influential when the traffic of Chinese indentured labourers were called "pig trade" and Chinese male immigrants were named "the gentlemen of the pigtails." The "pigtail" was the degraded metonymy not only for the subjective Chinese labourers but also for their emasculated masculinity in the American popular imagination, exemplified in a song "Big Long John" composed by Luke Schoolcraft (1873). In the song, a Chinese immigrant named John is attacked by a Native American who cuts off his queue while he is on the way to see "his sweetheart Chum Chum Fee." John further confesses to his lover that "he died from loss of his cue" (Schoolcraft, 1873, p. 13). The

Native American's act of cutting the Chinese male's queue deprives John's right to claim manhood, makes him shameful of the loss and leads to a failure to see his girlfriend, thus confirming Schoolcraft's portrayal of Chinese men as "less than manly" (ibid, p. 52). In this light, Louie revives the historic and imagined account of Chinese immigrants, reflected through the club ladies' further playful touches of Sterling's hair. Millie Boggs jokes that his ponytail will make "a delicious whip as she gives it a playful tug" (Louie, 2000, p. 40). Sharon Fox even treats the protagonist as a horse, grabs hold of his hair and calls out "Giddyup!" (ibid, p. 40). One may argue that the women's toying with his hair bears interesting similarities with Thomas Nast's (1869) famed portrait "Pacific Chivalry," in which an angry and hairy American named "California" with a whip uses Chinese ponytails for horse reins and prevents any Chinese escapes during the period of anti-Chinese sentiments. In this light, the women of the club make fun of the ponytail as an "authentic" phallic symbol of an exotic manhood and bring power to their suppressed femininity by claiming the role of "California," who threatens, controls and even kills the Chinese immigrants in order to reinforce their own physical and cultural superiority.

Under the American patrons' immense pressures produced by their pursuit of an "authentic" experience of Chinese culture and ethnicity in culinary and sexual terms, Sterling conjures an acute observation of his role in the entertainment of white elites:

> I can read [Libby] perfectly: Not only are the slides and the memories they hold hers, so are the people and objects in those pictures. And here I am, as if I'd just stepped off the screen, proof of her assertion.
>
> (Louie, 2000, p. 47)

Sterling's status as travel souvenir feeds the public's imperial nostalgia and becomes increasingly solidified during their discussion of Chinese body types. The women even devise a contest to guess the protagonist's height and figuratively take him as a prize for the winner. Hence, Sterling tragically becomes what he keeps refusing to cook – the tasty yet clearly exotic culinary delight.

The myth of egg foo yung

As the novel develops, Sterling's marriage to his Jewish girlfriend Bliss seems to relieve him from the discrimination received whilst working in the American club. Sterling experiences a brief moment of hope when Bliss's father Morton offers him a chance to host his own televised cooking show: "[Morton] wants to discuss the proposed cable show, he wants to put me and my telegenic hands on the air; such faith he has in me, Sterling Lung, the male Julia Child, the lean James Beard!" (Louie, 2000, p. 186). This fantasy of democratizing French cuisines in American households like those gastronomical gurus, however, is soon pulverised by Morton's disclosure of the cooking show's name, *Enter the Dragon Kitchen*, confirming the Jewish entrepreneur's appeal to Chinese exoticism for profits. Sterling continues to explain that the name is "a takeoff on the Bruce Lee film title, and an unequivocal pronouncement of the show's basic theme: Chinese cooking" (ibid, p. 204). Very much like the model of Bruce Lee, whose popularity allegedly symbolises a further marginalisation of Asian American immigrants during the 1970s (see Chan, 2002, pp. 73–96), Sterling's role in the show is also infused with the white audiences' alienation of Sterling's Chinese identity and the categorically Chinese cooking that accompanies it from the American characters. The Jewish father voices the essentialist association of the hero's Chinese-ness to the limited avenue of his culinary authority in the following conversation with Sterling:

> "Look, imagine you're a housewife, and you're looking to improve yourself . . . so you flip on the TV set for help. Who do want to teach you to crepe suzette, that fat James Beard or some Chinese guy?"
> "The Chinese guy who went to the CIA?"
> "No, Sterling. Not the Chinese guy," Morton Sass said. "But do your know what? That Chinese guy is where you go if you want to egg foo yung . . . You'll be a pioneer, Sterling. Like Columbus! Neil Armstrong!"
> (Louie, 2000, pp. 210–211)

On the surface, the asserted American housewives' reliance on the Chinese American chef to teach ways of cooking egg foo yung facilitates the dominant class's acquisition, accumulation and transmission of cultural capital through legitimizing the Western knowledge of exotic cultural artefacts. By labelling such an Americanised dish as egg foo yung as typical Chinese, the American diners give this culinary fusion a naturalness in a sense that fulfils their quest for exotic culture. In doing this, the dominant class gains control of what is originally in the hands of food producers and facilitates what Bourdieu (1984, p. 468) calls "the schemes of habitus" and "the primary forms of classification"—an obsession with difference that springs from the preservation of imperial knowledge and the ignorance of the other culture.

What Morton and the American housewives continue to overlook is the fact that the somewhat paradoxical logic behind the designation of exotic others grants liberating possibilities for the foreign cultures themselves. One way to understand this is to delve into the dynamics of "invented tradition" and "invention," the tropes that claim essential position in our understanding of culinary authenticity. Through a close reading of the term of invention in Eric Hobsbawm and Terence Ranger's (1983) *The Invention of Tradition* and Rebecca L. Spang's (2000) *The Invention of the Restaurant*, Andrew Warnes suggests an unnoticeable yet radical shift of meaning that has tended to accompany the transformation of understanding on inventions. While Hobsbawm presents "invented tradition" as a purposeful and political act in which the imposition of artificial truth inculcates a prior value and norm characterised by reference to the past and enacted by an often single and superior initiator, Spang, in observing the spread of restaurants in revolutionary France, dislodges invention from the hands of the authority and complicates the diverse cultures' command over the inventions in various ways including culinary creation (Warnes, 2013, pp. 352–353). Warnes (ibid, p. 354) sums up the transition:

> The concentration on *invention* comes at the price of a largely secret disinvestment in tradition. Its popularization is also a disarming, a dilution, which leaves the concept powerless to carry out its original work of spotlighting and exposing forms of cultural appropriation.

Hence the historical reinvention of barbecue in the American South, suggests the food critic, demonstrates that authenticity is achieved not by the pursuit of purity but by "the heritage of process and technique we recover from it" (ibid, p. 354).

Warnes' perception of culinary authenticity arguably invests new meanings into the invention of egg foo yung, the so-called Chinese cuisine in Morton's mind that achieves the Western fascination with Chinese food and race. The popularisation of these foods in America, especially during the first hundred years of its development, owed much to the versatility and adaptability of the Chinese restaurateurs that enabled them to cater to the sophisticated gourmets of Manhattan as well as less discriminating palates all over small American towns. Such efforts can be highlighted in the Chinese immigrants' efforts to organise their menus with startling homogeneity that can reflect a standard repertoire of tasty but bland Americanised versions of such Chinese dishes as chop suey, egg foo yung and bamboo shoots. In food historian Cynthia Ai-Fen Lee's words, the menu "has to be exotic enough that it's different, but they have to keep it familiar" (Lee, cited in Lou, 2004). In this light, the authenticity surrounding Chinese food grows from the hands of Chinese chefs, who are overwhelmingly governed by this gastronomic strategy that keeps Chinese food distant from and, in the meantime, attached to the American dietary expectation and familiarity. The immigrants' continuing investment in the possibility of integrated dishes situates them in the "third space" under Bhabha's ([1994] 2004) term, where the marginalised minorities use a third rhetorical space to disrupt and destabilise the central authority. Through the proactive initiation of exotic tastes and "invented tradition," the Chinese culinary inventions run counter to "the exoticism of multiculturalism" and achieve "the inscription and articulation of cultural hybridity" (ibid, p. 56). Therefore, the authenticity of egg foo yung that Morton buys into his U.S. hegemonic perspectives, borrowing Warnes' (2013, p. 359) analysis of terroir, "offers no dream of purity," and foreshadows Sterling's ability to renegotiate Chinese American identity in his performance on the cooking show.

Importantly, the chef infuses the political affects with his parody of Chinese culinary authenticity in the cooking show, especially during the shooting of one episode shortly after his son Ira's funeral. Sterling's introduction of "Shlimp and robster sauce" (Louie, 2000, p. 331) with an accent of Chinese pidgin is followed by his insightful contemplation of the dish's ingredients, among which salt draws his immediate attention. The chef then visualises such emotional sentiments by telling the audiences what he learns about salt:

The Jewish people used it in sacrifices and ceremonies. Homer described nations as poor whose citizens didn't mix salt with their food. The Roman Empire paid its soldiers a wage called *salarium*, or "salt money" (the root of our own *salary*). In the Middle Ages, the salt routes were the basis for the flow of trade. Across the centuries, state monopolies in the production and sale of salt brought wealth and power to nations and impoverished those so taxed. Gandi in 1930 as an act of civil disobedience scooped up a handful of salt crystallized from the evaporated waters of the Arabian Sea, symbolically breaking his British master's monopoly and striking a blow against the Empire.

(ibid, p. 332)

Sterling's presentation on an alternative cultural history of salt calls to the fore the notion of purity in that it confirms Sidney Mintz's (1996) belief in the complex workings in the creation of food's purity. Like what happens to the production of sugar and almonds, the purity of salt "is imputed to edible white substances" and is "processed to whiteness by human ingenuity" (Mintz, 1996, p. 90). In the protagonist's narration, the purity of the natural food substance builds upon not only the salt labourer's alteration, but also the regulating effects of the global markets in alienating food producers from salt as a valuable commodity. Salt has been crucial to the emergence of Western modernity not only as a commodity to be used as means of payment or routes of trade, but also as a byproduct of exploited labourer bodies. Both the improper ways of salt eating and the reliance of salt for empire expansion point to salt's pivotal role in national and international entities vying for economic power at the expense of the working classes who are marginalised and disempowered in the formations of world systems of trade, production and consumption.

However, as suggested in the example of Gandhi's use of salt as a symbolic tool during India's independence movement, the protagonist's sensitivity to salt defies the conventional inscription of salt as pure and natural. It is further constituted as an affront to civilised U.S. palates by investing this common ingredient with enriched figurations that arise not only from the personal sentiments of sadness or life, but also from remembering histories of civil disobedience against imperial dominance. With "each new bit of history remembered" (Louie, 2000, 332), Sterling's gustatory engagement with the working class's resistance empowers his impeded subjectivity set to feed the American audiences' exotic appetites, gradually strengthening the originally quivered voices and, more intriguingly, triggering collective memories of Chinese ancient ways of acquiring salt. In a sentimental account of something that he "never puts much stock in it," the protagonist tells the audiences that salt was invented by Chinese nearly three thousand years before the Western countries, making them the first to cultivate salt: "We flooded fields with seawater, and after its evaporation, we harvested the remaining crystals from the soil" (ibid, p. 333).

Sterling's transmission from remembering in the form of history to the collective memories of racialised coolie labourers, according to Lily Cho's (2005) analysis of the relationship between taste and memory, speaks to the enhancement of his diasporic subjectivity because these true memories recuperated from racialised history are often figured in collective gustatory desires, which pose radical challenges against U.S. imperialistic powers. Cho's belief in the power of such collective memories of tastes notably builds on Pierre Nora's differentiation between the history that binds a community and the memory that a community shares in its collectivity. She quotes Nora, "memory is by nature multiple and yet specific; collective; plural, and yet individual. History, on the other hand, belongs to everyone and to no one, whence its claim to universal authority" (Nora, quoted in Cho, 2005, p. 96). While Nora's differentiation between memory and history gestures against the singular tendency to historicise the ways to remember the past, Cho moves further to argue that the shared memory of a diasporic community is deeply embedded in the corporeality of the diasporic body, longing for authentic experiences that defy the binds of historicism. In this light, through an individual recuperation of the memory attesting to Chinese salt labourers from the history that justifies Western exploitation, Sterling "authenticates the possibility of a collective gustatory desire as a route to an alternative history" by "giving credence to a mode of knowing that has been suppressed and misnamed as sentiments" (Cho, 2005, p. 101). In his mind, the collective cultural memory that emerges from the unspoken but shared experience of exploitation and servitude mediates the gap between the past and the present and situates him "with the ancestors, among the world's first Lungs, smoothing the pans of seawater, pulverizing large sediments into edible grains" (Louie, 2000, p. 333). More intriguingly, Sterling's imagined integration into Chinese cultural roots gives birth to a collective ethnic community in which his father, the remaining son and himself "work the salt" (ibid, p. 333) together. With the deceased Ira, the three practice Chinese tradition to purify Ira's body with salt and get ready to protect others from calling him a barbarian.

Eating "Real" Chinese food

Towards the end of the novel, when Sterling faces the pressures of the divorce and his father's passing away, the hero manages to achieve the affective performance on the cooking show by identifying with the once-rejected Chinese heritage in a way that significantly points to a celebration of eating "real" Chinese food. He tells his son Moses that he knows "the perfect thing" to stimulate his appetites, which is "what my father used to make for me for a snack when I came home from school" (Louie, 2000, pp. 370–371). In his description, the concoction of saltine crackers, sweet condensed milk and boiled water becomes what he calls "comfort food, warming and soothing" (ibid, p. 371). The comfort of the culinary experience works to stir feelings of solace via nostalgia and to reconnect him to Moses through the resurrections of "good, safe, happy times" (ibid, p. 371) between him and his father. More intriguingly, the hero's growing ease with the cultural roots that his parents represent is followed by their discussion on the food's authenticity when they prepare the snack together. The young son casts doubt upon Sterling's claim that "it's Chinese," but it is immediately dismissed by the hero, who regards a pleasant father-and-son relationship as the key factor to determine the authenticity of the culinary creation. Sterling says,

> Let the steam caress your face, smell the roasted sweetness, the milk's own sugar, and feel the glow of well-being radiating from within . . . they are as real as the food is pure: just flour, water, sugar, milk, and salt.

Then, he tells Moses, "It really is Chinese, you know. Ah-Yeah used to make it for me. It's a special recipe he brought from China. And think about it, you and I just whipped this up together" (ibid, 372).

Here, Sterling enacts a new version of authenticity that is grounded in the father-and-son cooperation, one that bears the consciousness of ethnic heritage and guarantees the smooth inheritance of familial customs. Notably, the hero's labelling of the culinary blend of saltine crackers and condensed milk as "real" Chinese food negates the simplistic definition of culinary authenticity based on the food's origin. Instead, Sterling's sense of "real" is established upon the sentimental effects of any food and ingredients that are regarded "pure" enough to evoke traces of memory in remembrance of the deceased father and the lost cultural roots. In this light, through a schooling of his biracial son that seeks to transmit the renewed concept of culinary authenticity from one generation to another, Sterling engages in a material project of what Lisa Lowe (1996, p. 65) calls an "active cultural construction," the very effort that challenges a fixed profile of ethnic traits in a complex circle of hybridity and multiplicity. The hero's practice of cooking and eating food "that [is] partly inherited and partly modified, as well as partly invented" (ibid, p. 65) critiques the Western hegemonic values of otherness and questions the dominant conception of authenticity as fixed and singular. Through this, Sterling eventually comes up with a successful means to facilitate the making of a hyphenated identity that articulates the Asian American immigrants' ways of survival in the American society and is, significantly, deemed acceptable across the span of three generations.

In the final moment of the novel, Moses begins to understand the meanings of such practice and agrees with the mode of authenticity that his father espouses by claiming, "We just cooked Chinese food!" (Louie, 2000, p. 372). The call confirms the dynamic representations of eating "authentic" foods that defy U.S. hegemonic politics of defining ethnic characteristics as fixed and exotic. The cultural practice that originally reveals a Western desire to connect with a racist past during the Cold War turns into the one that constructs Asian American cultural identity and demonstrates a versatility similar to the characteristic of culinary authenticity. Sterling's imagined return to his Chinese homeland, in Louie's novel, is not portrayed as an easy connection with authentic cultural roots; however, it encounters an uncomfortable and complex negotiation between U.S. nationhood and ethnic identity as a performative arena to dismantle the notions of purity, origin and invention in the concept of authenticity. Eating and cooking "real" food framed in *The Barbarians are Coming* underscores the organic connections between places, people and items of consumption that transcend the boundaries of race, ethnicity, gender, history and memory. In other words, when pondering the novel's title, we, the readers, need to think anew about who the "real" barbarians are and what they are "truly" coming for, no matter in the literary space of the novel or in a realistic Chinatown restaurant.

References

Bhabha, H. ([1994] 2004) *The Location of Culture*. London: Routledge.
Bourdieu, P. (1984) *Distinction: A Social Critique of the Judgement of Taste*. Cambridge: Harvard University Press.
Bright, R. K. (2017) 'Migration, Masculinity, and Mastering the Queue: A Case of Chinese Scalping', *Journal of World History*, 28(3/4), pp. 551–586.
Chan, J. (2002) *Chinese American Masculinities: From Fu Manchu to Bruce Lee*. London: Routledge.
Cho, L. (2005) '"How Taste Remembers Life": Diasporic Memory and Community in Fred Wah's Poetry', in Koo, T. and Louie, K. (eds.) *Culture, Identity, Commodity: Diasporic Chinese Literature in English*. Hong Kong: Hong Kong University Press, pp. 81–106.
Heldke, L. (2003) *Exotic Appetites: Ruminations of a Food Adventurer*. London: Routledge.
Hobsbawm, E. and Ranger, T. (eds.) (1983) *The Invention of Tradition*. Cambridge: Cambridge University Press.
Huggan, G. (2001) *The Postcolonial Exotic: Marketing the Margins*. London: Routledge.
Lou, M. (2004) 'As All-American as Egg Foo Yong', *The New York Times*, 24 Sept. Available at: www.nytimes.com/2004/09/22/dining/as-allamerican-as-egg-foo-yong.html [Accessed 6 Mar. 2018].
Louie, D. W. (2000) *The Barbarians are Coming*. New York: G. P. Putnam's Sons.
Lowe, L. (1996) *Immigrant Acts: On Asian American Cultural Politics*. Durham: Duke University Press.
Mintz, S. (1996) *Tasting Food, Tasting Freedom: Excursions into Eating, Culture and the Past*. Boston: Beacon Press.
Nast, T. (1869) 'Pacific Chivalry', *Harper's Weekly*, 7 Aug. Available at: https://thomasnastcartoons.com/2014/02/25/pacific-chivalry-7/ [Accessed 26 Sept. 2018].
Padoongpatt, M. (2013) '"Oriental Cookery": Devouring Asian and Pacific Cuisine During the Cold War', in Ku, R., Manalansan, M. IV and Mannur, A. (eds.) *Eating Asian America: A Food Studies Reader*. New York: New York University, pp. 186–207.
Schoolcraft, L. (1873) *Luke Schoolcraft's Shine on Songster*. New York: A. J. Fisher.
Spang, R. L. (2000) *The Invention of the Restaurant: Paris and Modern Gastronomic Culture*. Cambridge: Harvard University Press.
Warnes, A. (2013) 'Edgeland Terroir: Authenticity and Invention in New Southern Foodways Strategy', in Edge, J. T., Engelhardt, E. S. D. and Ownby, T. (eds.) *The Larder: Food Studies Methods from the American South*. Athens: The University of Georgia Press, pp. 345–362.

4 Good authentic vibrations
The Beach Boys, California, and *Pet Sounds*

Christopher Kirkey

Introduction

As musician David Crosby (Fawcett, 1978, p. 14) stated, "If you want to look at the essence of California music, look at the Beach Boys." Between 1961 and 1965 popular music was principally characterized by songs featuring melodic arrangements and lyrics focused on the idyllic surf, sand, sun, summer, girls, and cars of Southern California. At the forefront of this new authentic place-based sound was one band – The Beach Boys. In song after song, album after album, The Beach Boys effectively crafted an original canon of music that was geographically specific and magically transported listeners around the world to the sun-soaked beaches and carefree lifestyle of Southern California.

By 1966, however, The Beach Boys would depart the world of Southern California to create an equally authentic yet non-place-specific sound. The Beach Boys' *Pet Sounds*, their twelfth album, released on May 16, 1966, is ranked by *The Times*, *NME*, and *Mojo* magazines as the greatest rock music album ever recorded and released, and is officially listed by the Library of Congress on the National Recording Registry as a seminal American musical contribution. The 1966 effort has been heralded by *Rolling Stone* magazine as the second-greatest album ever made, drawn from a list of 500 chart-topping titles.

Pet Sounds is an innovative masterpiece that not only permanently signaled a profound change in the sound of The Beach Boys, but also serves as an authentic benchmark in musical arrangement, featuring a fresh, unorthodox use and range of instrumentation, introspective lyrics, and innovative audio recording techniques – little to none of which today one would associate with the well-established California place-based vibrations that The Beach Boys popularized internationally. The entire album of thirteen songs (two of which, "Pet Sounds" and "Let's Go Away For Awhile," are instrumental compositions) was conceived of by The Beach Boys founder and leader Brian Wilson, with the assistance of lyricist, advertising executive, and first-time collaborator, Tony Asher.[1] Recording started January 18, 1966, largely at Western Recorders Studio 3 in Hollywood (four studios would be used in total, most especially Brian's favourite, Gold Star Studios in Los Angeles);[2] *Pet Sounds* represented a significant departure in the recording process and the now well-defined California sound for the Beach

Boys – Brian, Al, Carl Wilson, Dennis Wilson (Brian's two younger brothers), cousin Mike Love, and Bruce Johnston (Crowley, 2015, p. 23).[3]

This chapter, which is principally dedicated to examining the musical approach, recording history, critical reception, and impact of the *Pet Sounds* album, suggests that *Pet Sounds* is the most musically advanced recording undertaken by The Beach Boys, and arguably represents the greatest authentic American rock record ever created. The chapter focuses on a range of key questions and areas of investigation specific to the *Pet Sounds* album, including: (1) Brian Wilson's music and Tony Asher's words – how was *Pet Sounds* launched, what was the process behind the development of the album, and why did Brian elect to move away from the commercially successful California sound he had championed since 1961? In short, how did this collaborative arrangement come about, and how did Brain and Tony work together to craft the sound of *Pet Sounds*?; (2) why did Brian purposely choose to utilize arguably one of the most professional assemblage of studio musicians (known as The Wrecking Crew) to record all the musical arrangements (i.e., backing tracks) for each song, effectively almost completely foregoing non-vocal instrument contributions from The Beach Boys?; (3) a detailed evaluation – i.e., the authenticity of the musical arrangements, melodies, instruments utilized, chord progressions, vocal harmonies, and lyrics – of each of the thirteen songs; and, (4) the reception given to *Pet Sounds* upon its release, and an assessment of its long-term impact and place in rock music history.[4]

The chapter first turns to a careful examination on the musical history, production and evolution of The Beach Boys place-based California sound from September 1961 to May 1966.

Historical background: 1961–1965

In the summer of 1961, five young men (three brothers, a cousin, and a best friend) from Hawthorne, California would come together to form one of the world's greatest rock and roll bands of all times – The Beach Boys. Alan Jardine, a standup bassist, guitarist, and folk singer persuaded his friend Brian Wilson to form a singing group, which would include Brian's brothers, Carl and Dennis, and cousin Mike Love. At the suggestion of Dennis (the only surfer of the three Wilson brothers), Brian and Mike penned a song about surfing; the product of that collaboration, the song "Surfin'," became the focus of a non-stop Labor Day Weekend musical practice session (with Carl on guitar, Al on bass, Brian on snare drum, and lead vocals by Mike) by the newly founded band in the Wilson household garage. The Beach Boys' first recording session on September 15, 1961, took place in the Los Angeles Mayberry Street home studio of record producer (and acquaintance of Murry Wilson, father to Brian, Carl, and Dennis) Hite Morgan. Three songs were recorded that day: "Surfin'," "Luau," and "Lavender."[5] The only song of these three written and arranged by Brian Wilson and the band (then known, albeit briefly, as The Pendletones) was "Surfin'," which became a hit in the greater Los Angeles market and beyond.[6]

Throughout the balance of 1961 and 1962, the Beach Boys would continue to evolve and enjoy success as a band grounded in a distinct, original sound emphasizing rich harmonic vocals, original musical arrangements by Brian Wilson, and lyrical contributions focused on surf culture, sand, sun, summer, cars, and the girls of California.[7] The emerging surf sound of The Beach Boys, captured again in the 1962 hit single "Surfin' Safari," was significantly different than many of the surf bands then populating Southern California. Surf music was a decidedly popular albeit still a regional (if not local) rock and roll musical genre; key musical acts that featured surf music were overwhelmingly instrumental surf bands including Dick Dale and The Del Tones, the Surfaris, the Ventures, the Chantays, and the Challengers – all of whom focused on a dominant reverbed "wet" guitar sound designed to simulate the movement and crashing of ocean waves.[8] In simple terms, the sound of The Beach Boys was fuller, featuring a rich array of blended lead and backing vocals, and sported a more sophisticated musical arrangement. It was a new, distinct, authentic place-based sound focused on California.

The respective designation of musical assignments within the Beach Boys, both vocally and instrumentally, was refined through 1962: Mike and Brian would share and alternate lead vocals, Brian would concentrate on the Fender electric bass, Dennis on drums, Carl on a Fender Stratocaster (replacing his original Kay guitar), Al (to be briefly replace by Wilson neighbour, David Marks) on rhythm guitar, and Mike periodically on saxophone.[9] While a band in name, the principal creative force behind the music of The Beach Boys rested with one individual – Brian Wilson. A talented vocalist, arranger, and piano player from a very early age, Brian's formative musical influences rested most heavily on a singular musical vocal group, The Four Freshmen. A "clean-cut vocal quartet that sang old fashioned jazz-pop tunes with intricately arranged four-part vocal harmonies . . . it was the sound," Peter Ames Carlin writes, "of four dorky guys in sweaters that changed the course of Brian's life" (Carlin, 2006, p. 22).

Between 1962 and 1965, the Brian Wilson-led Beach Boys would make and release no less than eleven albums. Each offering featured a mixture of Brain Wilson original musical compositions (with words most often provided by an ever-rotating lineup of lyricists including Mike Love, Gary Usher, and Roger Christian), instrumentals, and cover-songs. Major hits for The Beach Boys in this period included "Surfin' Safari," "Surfin U.S.A.," "Fun, Fun, Fun," "I Get Around," "Help Me Rhonda," and "California Girls." Brian Wilson had also taken on the role of exclusive record producer for The Beach Boys by late 1962 (a task wrestled away from The Beach Boys' label, Capitol Records, by Murry Wilson), effectively ensuring that all musical production would bear Brian's final imprint – as a musician, vocalist, arranger, sound mixer, and "unofficial" audio recording engineer.

The song "Surfin' U.S.A." would prove to be the foundational cornerstone upon which The Beach Boys' California sound was crafted. According to music writer Mark Dillon (2012, p. 1), "Surfin' U.S.A" "helped take surfing from a Southern California craze to something kids all over the country could do – if they only had an ocean." In November of 1962, Brian Wilson – directly inspired by R&B sound of Chuck Berry's 1958 song "Sweet Little Sixteen" – began to compose a "new" surf song; a song that would feature Berry's melody (albeit with a fresh, new musical arrangement) and new lyrics dedicated to a listing of popular surfing areas in the United States and beyond. Brian first worked out the musical arrangement for the song on the upright piano in the Wilson family music room.[10] Brian then approached David Marks (playing a Fender Stratocaster) and his brother Carl (who had "recently swapped his Stratocaster for an Olympic white Fender Jaguar, a guitar with a shorter neck and more distinctive sound") to create the guitar sound – especially the opening sequence and the solo by Carl in the middle – for the song (Crowley, 2015, p. 75). Drumming duties on "Surfin' U.S.A." would be shared by Dennis Wilson and Frank DeVito, while Brian would play the Hammond organ and bass. Lyrics for the song – a list of the hottest surfing locations – were provided at Brian's request, "by a surfer named Jimmy Bowles – the baby brother of his new girlfriend, Judy Bowles" (Carlin, 2006, p. 35). As Brian would later acknowledge, "Jimmy was a surfer. I asked him to make a list of every surf spot he knew, and by God he didn't leave one out" (Wilson and Gold, 1991, p. 67).[11]

During recording sessions, Brian hit on the idea of double-tracking the lead vocal of Mike Love – i.e., effectively layering two separate, but nearly identical vocal performances one on top of the other to create the lush, full sound of the song. This was the first of many Beach Boys songs to use this technique; a technique, Carlin notes, that produced "a fullness and brightness that made it leap out of the speaker" (Carlin, 2006, p. 41). The four choruses of "Surfin' U.S.A." also feature background vocals and were recorded over the course of eight and one-half hours during three sessions on January 5 (Western Recorders studio, Hollywood), January 16 (Fine studio, New York), and January 31 (back at Western) – (Badman, 2004, p. 32). Performed live for the first time on February 2, 1963 at a March of Dimes fundraiser in San Bernardino, California, and released on vinyl March 4, 1963, "'Surfin' U.S.A.' became a massive hit. It dominated U.S. radio during the spring and summer of 1963, and stayed in the national Top 40 for four months" (Rusten and Stebbins, 2013, p. 18; Stebbins, 2011, p. 34). As David Leaf observes, "Surfin' U.S.A." was "a surf anthem . . . that extolled the virtues of the surf life . . . Brian struck a truly responsive chord. His universal surfer hit home in even the most land-locked locales" (Leaf, 1978, p. 40).

"Surfin' U.S.A." would be proof to other musical vocal surf performers, such as Jan and Dean, Bruce & Terry, The Rip Chords, First Class, and Ronny & the Daytonas, that pop songs built around and promoting place-based Southern California surf culture could be internationally successful.[12] "'Surfin' U.S.A.,'" Jadey O'Regan writes, "turned the idea of surfing, an established lifestyle experienced primarily on the West Coast, into a more accessible and inclusive preoccupation" (O'Regan, 2016, p. 146). Ironically, perhaps the success of "Surfin' U.S.A." was most directly felt on The Beach Boys themselves – through 1965, successive songs focusing on surf culture, such as "Catch a Wave" and "Girls on the Beach," and rockin' Chuck Berry-like guitar numbers, as featured in the song "Fun, Fun, Fun," remained at the centre of The Beach Boys' recording and performing universe.

The Beach Boys, argues Daniel Harrison, were "almost single-handedly responsible for a national surf-music fad in the early 1960s" (Harrison, 1997, p. 34). As much as The Beach Boys' music was principally responsible for internationally popularizing surf culture, their sound, above all, manufactured, promoted, and mythologized an inviting, arresting vision of Southern California; a vision dominated by a fantasy grounded in a carefree sun-soaked lifestyle focused on sand, surf, fun, girls, and fast cars. "The surfing lifestyle, most notably in southern California," observes Kirse Granat May in her work, *Golden State, Golden Youth: The California Image in Popular Culture, 1955–1966*, "spawned a music that schooled teens across the country in the sport's appeal, offering a teenage element of the California dream . . . creating a mythic place of carefree consumption and endless summer" (May, 2002, p. 98). As Larry Starr and Christopher Waterman (2014) underscore in their work, *American Popular Music*, the sun and surf of Southern California

> was celebrated in song after song by the Beach Boys. These songs indelibly enshrined [Brian] Wilson's somewhat mythical vision of California in the consciousness of young Americans – to such an extent that still, for legions of pop music fans, merely the titles are sufficient to summon an entire state of mind . . . Wilson's vision was appealingly inclusive, even as it remained place-specific.
>
> (Starr and Waterman, 2014, p. 300)

"The Beach Boys could not help but mythologize a landscape and way of life," Kevin Starr (2009) observes

> that was already so surreal, so proto-mythic, in its setting. Cars and the beach, surfing, the California Girl, all this fused in the alembic of youth: Here was a way of life, an iconography, already half-released into the chords and multiple tracks of a new sound. The songs of the Beach Boys upgraded an overnight place into national identity by connecting its myth to young America and inviting young America to buy into the dream, to find its own Southern California, if only within itself, wherever radios were playing, and whatever cars were cruising down whatever streets.
>
> (Starr, 2009, p. 373)

While The Beach Boys popularized the sound of Southern California, an examination of their musical canon between 1961 and 1965 clearly illustrates that Brian was also interested in crafting songs that were non-placed based, and instead were deeply personal and introspective in nature. These private statements can be most clearly be found in songs like "In My Room," a fabled reference to the music room in the Wilson home where Brian would spend so much time. "It served as a sanctuary to me . . . I slept there, right beside the piano" (Lambert, 2007,

p. 97). 1965's "Please Let Me Wonder" is a direct outpouring of Brian Wilson's emotions; it is arguably his most heartfelt and autobiographical composition.[13] Consider the revealing nature of the narrative outlined in the following verses:

Now here we are together

This would've been worth waiting forever
I always knew it'd feel this way

And please forgive my shaking
Can't you tell my heart is breaking?
Can make myself say what I planned to say
I built all my goals around you
That some day my love would surround you
You'll never know what we've been through

For so long I thought about it
And now I just can't live without it
This beautiful image I have of you.[14]

In addition, the music of The Beach Boys by mid-1964 and into early 1965 demonstrates an increasing range of musical complexity, as measured by melodic arrangements, range and types of instrumentation, and chord progressions. The twenty-two-second instrumental (highlighted by a slower tempo than the rest of the song) opening to 1965's "California Girls" features 12-string guitar chords played in chamber echo, electric base, cymbals, French horn, and saxophones. In October 1965, The Beach Boys recorded "The Little Girl I Once Knew," featuring yet more notable hints in the evolution of the band's sound.[15] The song's opening, most obvious for the 12-string guitar and pulsating organ, leads to verses unexpectedly culminating in "an unusual four-beat stop, sustaining the last note." (Dillon, 2012, p.72). After a delay, the chorus starts up. This song structure, with a built-in pause or respite of almost two seconds between verse and chorus (repeated twice), was novel for pop music. Yet another song offering that expanded the musical range of The Beach Boys, found on *The Beach Boys Today!* album, is "In The Back of My Mind." More than any other song recorded by The Beach Boys, this composition is most closely linked to the sound later featured on *Pet Sounds*. Musically, it is a breakthrough record, featuring "chord progressions . . . virtually unprecedented in Brian's previous work . . . rhythmically, melodically, harmonically, and instrumentally, 'In the Back of My Mind' is like nothing Brian had done up to that date" (Lambert, 2007, pp. 190, 191). "Much of the work, particularly on the B side of the album *Beach Boys Today!*," Bruce Johnston observes, "departed from the scope of the Beach Boys previous sound" (Lambert, 2007, pp. 190, 191).

By late 1964, Brian Wilson formally withdrew – following a severe anxiety attack on December 23 while on a plane en route to Houston, Texas – from regular concert performances with The Beach Boys; his full-time focus would now be on creating records in the studio. "You guys will tour, I will stay in the studio and make music," Wilson told his bandmates (Lambert, 2007, p. 175). Beach Boy Bruce Johnston underscores the multiplicity of tasks (and related pressures) that Brian Wilson was uniquely shouldering on behalf of the entire band:

> Brain had all these jobs, he had to melodically write it, arrange it, go produce the tracks, vocally arrange what everybody sang, sing often the high lead part all these jobs and then go on the road . . . he now did the writing, arranging, producing, and recording – he was in charge of all facets.[16]

Brian was now working, in the words of music scholar Philip Lambert (2007),

> in the new comfort of self-imposed creative exile. His primary means of musical communication, his singing group that he had worked so hard to mold and develop and support in concert with his own vocal gifts, was now solely another component of his studio creations, musicians with roles to play not unlike the gifted instrumentalists who would provide the backing sounds.
> (Lambert, 2007, p. 194)

Pet Sounds: A new, equally authentic sound for The Beach Boys

The release of the Beatles' *Rubber Soul* in 1965 further expanded the frontiers of popular music, suggesting than an album could represent a complete work beyond simply being a vehicle showcase for one, two, or even three hit singles. *Rubber Soul* suggested that it was in fact possible to make a statement; an album could be conceived of and recorded as a coherent, thematically unified body of work. As *Rolling Stone* editor Anthony Decurtis notes, the idea of a record album not

> hanging on a hit single was revolutionary . . . so when Brian Wilson heard *Rubber Soul*, what he heard was a suite of songs . . . Brian could now conceive of an album consistent in sonic identity and lyrical narrative.[17]

Pet Sounds was a clear departure from the traditional California place-based musical arrangements and lyrical content that had come to dominate The Beach Boys sound since 1961. Brian Wilson would in fact produce a concept album in which each song would not only be thematically related to and unified by a common narrative storyline, but would also feature "an over-arching unity of musical sound and detail" (Lambert, 2007, p. 250).[18] "It was," one observer writes, "about a different formulation of the Beach Boys sound" (Lambert, 2007, p. 221). *Pet Sounds* sounded like nothing Brian and his bandmates had ever produced; the album was clearly *rooted in* California, but the musical and lyrical content suggested, for the first time in their careers, that the defining Beach Boys sound was *no longer place-based in the Southern California experiences* of the early-mid 1960s; a place that the worldwide general public had become accustomed and attached to. Profoundly pensive, self-reflective, and personalized, the album is a holistic musical treatment of romance in all its stages: the sequencing of the songs, for example, clearly suggests a unified narrative beginning with romantic innocence and guarded optimism ("Wouldn't It Be Nice," "You Still Believe in Me"), to uncertainty ("Here Today," "Don't Talk (Put Your Head on My Shoulder)," "That's Not Me," and "God Only Knows"), to disillusionment ("I Just Wasn't Made for These Times," and "Caroline No"). It was, in the view of Peter Ames Carlin:

> essentially a concept about the arc of a love affair from "Wouldn't It Be Nice," the sweetness and innocence of that, all the way to the end with "Caroline No," where he's looking at this older, harsher version of the young woman he met at the beginning of the album . . . The songs are about living away from home, trying to invent your own character, songs about this deepening love affair.
>
> (Brian Wilson Songwriter, 1962–1969, 2010)

Brian's principal collaborator in writing lyrical content for *Pet Sounds* would prove to be Tony Asher, a copywriter at the Carson/Roberts advertising agency in Hollywood. Referred to Brian by mutual friend Loren Schwartz, Asher had no previous experience with writing pop song lyrics. The collaborative process developed between Wilson and Asher was relatively straightforward; the latter would typically arrive at the Wilson Laurel Avenue home in Beverly Hills around noon and spend an hour or so with Brian enjoying lunch or simply having a coffee (Granata, 2003, pp. 82, 83). Asher notes,

> We would generally reminisce about our girlfriends in high school: the ones who broke our hearts and the ones that turned out to be very different than we had imagined . . . we'd set a mood and then go write a song influenced by that conversation.
>
> (Dillon, 2012, p. 90)

Brian Wilson would work on the musical arrangements for a song and Asher would take the lead, with input from Wilson, on specific lyrical elements. "The general tenor of the lyrics was always his [Brian's]," observes Asher, "I was really just his interpreter" (Leaf, 1990, p. 8). At times, Brian would provide Tony with a draft, and in some cases the completed backing track, to take home to pen lyrics for. Over a three-week span in January 1966, eight compositions would emerge that serve as the core of *Pet Sounds* (Wilson would add an additional five songs between January–April to complete the album).

The recording of the album, Brian determined, would be carried out in two distinct phases. Working at four different recording studios in the greater Los Angeles area, Wilson first sought to identify and secure the services of the most widely respected professional assemblage of studio musicians, known as The Wrecking Crew, to record all the musical arrangements (i.e., backing tracks) for each song. This decision was ultimately made easier as The Beach Boys were currently touring in Japan. With the assistance of audio engineers Chuck Britz and Winston Chong, upwards of twenty musicians would gather, working to transform Brian Wilson's ideas into music. The recording process typically featured three or four tracks that would be blended into a mono master on one dedicated track of an eight-track machine, effectively leaving the seven remaining tracks for vocals. Wilson personally oversaw and dispensed a constant range of instructions to each and every individual, seeking precisely the right tone, tempo, chord, or key. Being a formative part of Brian Wilson's creative process, Kent Hartman suggests, was challenging, "for three grueling months, Glen Campbell, Carol Kaye, Hal Blaine, and the other musicians toiled on *Pet Sounds*, sometimes working in the studio from seven at night until early the next morning – often on just one song" (Hartman, 2012, p. 154). Brian employed

> unconventional combinations of instruments to produce new and exotic sounds. In addition to having multiple musicians playing guitar, bass, and keyboard parts simultaneously . . . he mixed conventional rock instrumentation with various exotic stringed instruments, theremin, flutes, harpsichord, bicycle bells, beverage bottles, and even the barking of his dogs Banana and Louie.
>
> (Schinder, 2008, p. 114)

The second phase of recording took place as Brian worked with members of the band to learn and ultimately lay down the complex vocal harmonies designed to match the equally innovative backing tracks. *Pet Sounds* "marked the first full album where the Beach Boys' participation was limited almost exclusively to recording vocals . . . with *Pet Sounds* they [all the members of the band minus Brian] saw themselves simply as singers" (Crowley, 2015, p. 156). Mike Love characterized the vocal recording sessions as demanding:

> We worked and worked on the harmonies, and if there was the slightest little hint of a sharp or a flat [note], it wouldn't go on. We would do it over again until it was right . . . Every voice had to be right, every voice and its resonance and tonality had to be right. The timing had to be right. The timbre of the voices just had to be correct, according to how he (i.e., Brian) felt. And then he might, the next day, completely throw that out and we might have to do it over again.
>
> (Crowley, 2015, p. 156)

Pet Sounds: the album

Pet Sounds features thirteen songs; what now follows is a comprehensive review of the key features and a brief discussion of the musical and lyrical elements of each song title.[19]

The focus for the first song, "Wouldn't it be Nice," according to Brian Wilson, came from his infatuation with his wife Marilyn's sister, Diane Rovell. Brian provided Tony Asher with a tape of the instrumental backing track, and Asher went home and returned the next day with lyrics in hand (Mike Love would later add the lullaby-like taglines). The suggestive song is pitched from the perspective of an adolescent couple yearning for a true adult-style romance, including marriage ("then we'd be happy"). It is, in the words of Zooey Deschanel, "a grand statement about the unwavering optimism of young love" (Dillon, 2012, p. 109). The song features, for 1966, some unexpected, unorthodox choices in instrumentation including the use of two accordions which give the song an upbeat tempo, and a detuned 12-string guitar run directly into the recording console with live reverb added. The surprise tempo change in the ritardando, immediately before the bridge, slows the pace down dramatically.[20]

In the song "You Still Believe in Me," Brian effectively created an innovative way to use the scale, as his progressions are always ascending, "then pausing before they go up again" (Leaf, 1990, p. 9). As Tony Asher notes, the unique sound at the opening of this song can be traced to Brian's determination to identify precisely the right sound. This meant, in Asher's words, that "one of us [Brian] had to get inside the piano to pluck the strings, while the other guy [Tony] had to be at the keyboard pushing the notes so that they would ring" (Leaf, 1990, p. 9).[21]

"That's Not Me" is a journey in self-examination and maturity; the young man featured in this song has gone through "all kinds of changes" and is determined to "prove that I can make it alone." Effectively a search for identity, the song features a dramatic key change – from A major at the outset, concluding in F-sharp major. Earlier on during the day the backing track and vocals were recorded, The Beach Boys posed with a variety of animals at the San Diego Zoo; pictures that would form the basis for the front and back cover of *Pet Sounds*.[22]

The ethereal, dream-like feel (with vocals to match) of "Don't Talk (Put Your Head on My Shoulders)" opens with a bass-line introduction and features, by all standards, an "intensely chromatic verse chord progression" (Lambert, 2007, p. 228). With no backing vocals, Brian's brooding voice is double-tracked as he sings about the theme of unspoken love (a theme that would reemerge in the song "Good Vibrations").[23]

The fifth song on *Pet Sounds*, "I'm Waiting for the Day," traces its genesis back to 1964, when Brian Wilson copyrighted the recording as a solo effort. The musical arrangement was updated by Brian and squarely fits within the musical approach that characterizes *Pet Sounds*. Featuring an English horn in the first verse, a break for flute solos, and a lush string accompaniment to close the song, "I'm Waiting For The Day" narratively focuses on compassion; compassion and support for a young girl who has just had her heart broken. The final message of the song is to reassure the girl that her new love interest will not similarly hurt her, with Brian singing, "He hurt you then, but that's all done/ I guess I'm saying that you're the only one."[24]

Interviewed in 1967, Brian Wilson exclaimed that "Let's Go Away For Awhile," stands as the

> most satisfying piece of music I've ever made. I applied a certain set of dynamics through the arrangement and the mixing and got a full musical

extension of what I'd planned . . . I think the chord changes are very special. I used a lot of musicians on the track – twelve violins, piano, four saxes, oboe, vibes, a guitar with a coke bottle on the strings for a semi-steel guitar effect.
(Leaf, 1990, p. 10)

Tony Asher in fact wrote lyrics to this song, as per Brian's request, but Wilson ultimately chose to move forward with only the instrumental version. The vibraphone launches the song, featuring two basses, four saxophones, and a large string section building in tempo, and incorporates a bridge with an oboe and the aforementioned coke-bottle guitar effect.[25]

Capitol Records, concerned that the twelve-song compilation created for *Pet Sounds* lacked a potential hit single consistent with the California place-based sound long associated with The Beach Boys, opted to include "Sloop John B," a song that was recorded in the summer/fall of 1965, and not intended for inclusion on the album. The acapella vocal break had not previously been featured in a pop record prior to 1966. A traditional song dating to at least 1916, it was the 1958 Kingston Trio version that resonated with Alan Jardine.[26] The introduction to the song, featuring flutes by Steve Douglas and Jim Horn, sets a light and upbeat mood. Many observers, given the fact that the song remains distinctly outside the unifying conceptual framework of the album, have long suggested that the inclusion of "Sloop John B" is misplaced. As The Beach Boys' Bruce Johnston put it, "that's the fly in the ointment on *Pet Sounds* . . . it's brilliant, but it doesn't make any sense being on that album."[27]

Twenty or so musicians recorded "God Only Knows" live at Western Studio 3, which features an opening and closing French horn, dual accordions, a four-note flute theme, sleigh bells, an inverted bass line, a string quartet, and closing drum triplets. The song features a lead vocal by *nineteen-year-old* Carl Wilson; his voice on this track in uniformly described as "angelic." The harmonic segment that closes out the song, featuring Brian, Bruce Johnston, and Carl, is made even richer by Brian Wilson's overdubbing using Columbia Studios' 8-track recorder. The instrumental break, which is played staccato, was suggested to Brian – who was having difficulty getting the bridge – by session pianist Don Randi. Instrumentation for this song featured a prominent French horn, sleigh bells, a clip-clop percussion, a harpsichord, flutes, bass clarinet, double accordions, and strings. The song features a nicely crafted chord progression. Indeed, it's unclear as to what key the song is in; it initially sounds like E at the outset, then it sounds like A. In short, "God Only Knows" never really stabilizes on a particular key. The lyrics squarely contemplate a hopeless world in which a young man loses the love of his life. How is he to carry on? In Asher's perspective, if the girl chooses to end the relationship, the world will not suddenly cease to exist, but for the affected young man, he doesn't see the point in continuing to live. The song was first publicly performed by The Beach Boys on Thursday, July 28, 1966 at a concert in Massachusetts.[28]

Terry Sachen, The Beach Boys' road manager, co-wrote the bulk of the lyrics to the next song on *Pet Sounds*, "I Know There's an Answer," along with Brian (Mike Love later added small revisions to the text) in late 1965. Given the working title "Hang On To Your Ego," by Brain, the song once again speaks to a young man who, in the process of maturing, is in search of his identity: "I know there's an answer/I know now but I have to find it by myself." The central message of the song is self-assurance and affirmation; the young man can in fact successfully navigate the waters of life on his own. Framed by an instrumental opening and closing, the sound texture of the song is incredibly rich.[29]

"Here Today" features innovative tempo changes, horn arrangements, and a deep bass sound. During the instrumental bridge, one can hear Brian provide directions to audio engineer Chuck Britz; sounds that Brian ultimately chose to leave on the final version of the song. The lyrical focus provided by Tony Asher offers words of caution (or wisdom) from a jilted suitor to his would-be successor, suggesting to him that the girl of his dreams is not all that she appears to be, and that things could quickly change for the worse. Brian, at the suggestion of saxophonist Steve Douglas, purposely opted to record the backing track for this song at Sunset Sound in order to utilize the studio's distinctive echo chamber. The echoed effect – most clearly heard in the sound of the Hammond organ, along with the electric bass being played in an octave higher than usual, the use of five saxophones and marching trombones – creates a sound tailored to Asher's lyrics suggestive of a volatile romance.[30]

"I Just Wasn't Made for These Times," arguably the most personalized and introspective track on *Pet Sounds*, features the first use of the theremin by The Beach Boys (it would be featured prominently in the song "Good Vibrations").[31] According to Tony Asher,

> Neither one of us was a particularly popular kid . . . I think Brian was always very shy. Certainly I was not a guy all the girls wanted to go out with. So we talked about feeling that we were not part of the in-crowd when we were in high school. That's how that song got started.
>
> (Dillon, 2012, p. 92)[32]

The title track of the album, the song "Pet Sounds," clearly references Brian's personal preference for a new direction in the sound of the Beach Boys; something more instrumentally complex and simultaneously introspective in outlook.[33]

The final contribution to *Pet Sounds*, "Caroline No," was principally penned by Tony Asher, and were in the lyricist's view, "written because Brian was saddened to see how sweet little girls turned out to be kind of bitchy hardened adults" (Leaf, 1990, p. 15). In a conversation between Asher and Wilson, they lamented the memory of a sweet but now distant girlfriend; time had passed, and now when they encountered each other some years later, she had clearly changed. The lyrical content of the song is imbued with a forlorn feeling of what was. As Brian sings, "where did your long hair go, where is the little girl I used to know, how could you change that happy glow." "Caroline No" famously features Wilson's two dogs barking – Banana, a Beagle and Louie, a Weimaraner. Recorded in March at Western Studios, Brian blended their "pet sounds" as the closing feature of the song. The unique percussive effect that opens this song is attributable to Hal Blaine playing an upside-down, empty Sparkletts water bottle.[34]

As accomplished as *Pet Sounds* is, one song in particular, introduced and worked on during the writing and recording sessions for the album – indeed arguably the greatest single composition of The Beach Boys – was purposely not chosen by Brian to be included; namely, "Good Vibrations." Why wasn't this progressive song included on the album? According to Alan Jardine, "At that time, we had all assumed that 'Good Vibrations' was going to be on the [*Pet Sounds*] album . . . but Brian decided to hold it out. It was a judgment call on his part" (Crowley, 2015, p. 153). Brian Wilson's musical arrangement and Tony Asher's lyrics focusing on the "unspoken communication in boy-girl attraction" (to be later significantly updated by Mike Love) were, as Jardine suggests, not yet ready according to Brian (Dillon, 2012, p. 124). He wanted additional time to complete the song.[35]

Short-term reception and long-term impact

On May 16, 1966, The Beach Boys' *Pet Sounds* was released in the United States, with little enthusiasm or marketing panache, by Capitol Records (global distribution would come some six weeks later). Capitol had little interest in, and even less enthusiasm for, a Beach Boys album that sought to break new musical ground: executives at Capitol's Hollywood headquarters wanted an album littered with hits celebrating, once again, the familiar placed-based southern California themes of surf, sun, sand, summer, cars, and girls. *Pet Sounds*, viewed as unnecessarily musically complicated and almost devoid of any commercially oriented songs, was released in two audio formats: "monaural and a faux stereo called 'Duophonic,' in which the lowers registers of a monaural recording were emphasized on one channel while the higher registers were emphasized on the other" (Crowley, 2015, p. 155).[36] Capitol Records' dissatisfaction with *Pet Sounds* was made perfectly clear, when the company quickly – some seven weeks after *Pet Sounds* – released a Beach Boys greatest hits album, *Best of The Beach Boys*, squarely focused on the California sound that the Beach Boys had so successfully popularized.

As Kent Hartman (2012) observes about the release of *Pet Sounds*:

> a funny thing happened once this future classic made its way to the nation's record stores: it landed with a resounding thud. A confounded public just didn't know what to make of all the melancholy and introspection. Was this really the Beach Boys? The same fun-loving band that used to sing about waxing down their surfboards and liking California girls the best?
> (Hartman, 2012, p. 155)

Despite the change in the sound of The Beach Boys with *Pet Sounds*, it would be an overstatement to suggest that the album was in any way a commercial failure. "Let's just say," notes Jon Stebbins (2011), "it didn't quite meet the standard the Beach Boys were used to. The band's previous five LPs had charted at #4, #1, #2, and #6 respectively in the U.S., while *Pet Sounds* only made it to #10." (Stebbins, 2011, p. 71). In addition, the album remains the only Beach Boys production to produce four Top Forty singles, including "Sloop John B," "Caroline No," "Wouldn't It Be Nice," and "God Only Knows."

In 1997, Capitol Records released what would prove to be a GRAMMY-nominated box set, *The Pet Sounds Sessions*, featuring an examination of the making of the album, the first stereo mix of the album, complete backing tracks and vocal tracks, and numerous outtakes from the recording sessions.[37] *Pet Sounds* was given a platinum certification in 2000, and in 2001 an All-Star Tribute to Brian Wilson at New York's Radio City Music Hall was held, which included performances by various musicians (e.g., Elton John) of each song on *Pet Sounds*. From 2000 through 2002, Brian Wilson went on a world tour playing for the first time ever, the entire *Pet Sounds* album live in concert.[38]

Pet Sounds, as a concept album, also proved to be highly influential on a generation of rock and roll artists throughout the 1960s and 1970s. Perhaps the most obvious example is The Beatles' release of *Sgt. Pepper's Lonely Hearts Club Band* in 1967. Other significant musical acts who championed concept albums include the Moody Blues' *Days of Future Passed* (1967), Yes', *The Yes Album* (1971), Pink Floyd's *Dark Side of the Moon* (1973), and Supertramp's *Crime of the Century* (1974).[39]

Having recently marked the fiftieth anniversary release of the album *Pet Sounds*, Philip Lambert (2007) persuasively writes:

> In the decades since, *Pet Sounds* has only grown in stature and universal acclaim. It regularly appears at or near the top of "greatest album ever" lists and is fully recognized as an extraordinary accomplishment with far-reaching impact on subsequent popular music ... *Pet Sounds* is indeed a work of art of the highest caliber, a remarkable and radical rethinking of what an "album" is and can be.
>
> (Lambert, 2007, p. 223)

Conclusion

With *Pet Sounds*, The Beach Boys introduced the world to a genuinely authentic and unexpected sonic experience; a sound that was unquestionably a marked departure from the musical foundations they had crafted between 1961 and 1965. Those very foundations, tethered in the sun, sand, and surf of Southern California, are equally original and decidedly transformative. Popular music lovers around the world continue to be, through the sound of The Beach Boys, acoustically and visually drawn to a hedonistic, carefree place where unparalleled sun-soaked fun rules the day. The very identity of Southern California will remain forever linked to the musical contributions of The Beach Boys.

Notes

1 The sole exception to this is the song "Sloop John B.," popularized as a folk song by The Kingston Trio. The musical arrangement for the *Pet Sounds* version of the song was created by Brian Wilson. Long a favourite of bandmate Alan Jardine's, The Beach Boys in fact recorded this song over 14 takes, on July 12, 1965 – long before Brian began to work on the musical arrangements for *Pet Sounds* in late 1965. Jardine successfully lobbied Brian to include the song on the 1966 album.

2 The final recording session for *Pet Sounds* focused on a second vocal overdubbing for the song "I Just Wasn't Made for These Times," and took place at Columbia studio (A) on April 13, 1966. Gold Star Recording Studios, founded by David S. Gold, was located in Hollywood at 6252 Santa Monica Boulevard at the intersection of Vine Street.
3 Johnston would ultimately join the band as an official member of The Beach Boys in 1965, following Brian Wilson's first significant mental health episode (diagnosed at the time as an anxiety attack and nervous breakdown while on a flight from Los Angeles to Houston on December 23, 1964). It should be noted that Brian's immediate full-time replacement for live stage performances (as well as continuing in the capacity of recording session guitarist), was Glen Campbell. Johnston's first involvement with The Beach Boys would come on March 30, 1965 as an organist during the recording session for the song "Sandy."
4 The sole exception on *Pet Sounds* would prove to be the song "That's Not Me," featuring contributions from all three Wilson brothers.
5 Those initial recordings were released in 2016 as part of The Beach Boys, *Becoming the Beach Boys: The Complete Hite & Dorinda Morgan Sessions* (Los Angeles: Omnivore Recordings, OVCD-186, 2016).
6 "Surfin'" was released on record label Candix 331 on November 27, 1961. According to David Leaf's, 1978 work, *The Beach Boys and the California Myth*, "Surfin' made it into the top three in the L.A. radio charts, and peaked in *Billboard* magazine's national charts at 75. The record sold somewhere between ten and fifty thousand copies, and the group received a check for less than a thousand dollars" (Leaf, 1978, p. 32). For the most comprehensive, detailed account of The Beach Boys' musical activities (i.e., songwriting, recording, and performances) between 1961 and 1963, see Murphy, 2015.
7 The Beach Boys' first public performance as a band, during which they played two songs, took place at the Rendezvous Ballroom, 608 East Ocean Front, Balboa Peninsula in Newport Beach, California, on December 23, 1961 (Murphy, 2015, p. 343).
8 Dick Dale and the Del-Tones, the principal instrumental surf band, "released two full-length albums, *Surfer's Choice* (1962) and *King of the Surf Guitar* (1963), which brought his music to a national audience" (Tuttle et al, p. 358). As Kirse Granat May notes, "The surf music craze gained momentum, yet Dale never achieved a national following to match his local appeal. An East Coast tour in the summer of 1963 failed to attract the attention the record company had hoped, despite television appearances, profiles in *Life* and *Newsweek*, and persistent advertising by Capitol. Dale sang in rhythm and blues baritone, a sound that clashed with the youthful California image. Perhaps he was too old, or his instrumental guitar music lacked the lyrics to create a fuller picture of teen life." (May, 2002, pp. 101–102).
9 For a comprehensive list of the instruments owned and used by the entire Beach Boys lineup, please consult Stebbins, 2011, pp. 288–291.
10 The very first recorded version of "Surfin' U.S.A.," featuring only Brian on the piano and a completely different set of lyrics, can be found on The Beach Boys, *Hawthorne, CA*, 2-CD set (Hollywood: Capitol Records, CDP 531583, 2001).
11 Wilson and Gold, 1991, p. 67. Wilson's 2016 autobiography effectively makes the same observation: "[Jimmy] was always talking about the points and the spots where the great surfers wanted to go. I knew that I wanted to do a Chuck berry-type song about surfing, and whenever Jimmy talked about surfing I liked the names of the places. I asked him for a list" (Wilson and Greenman, 2016, p. 119).
12 Non California-placed musical bands that adopted and featured a surf/California sound are discussed in May, 2002, pp. 110–111.
13 Some music historians suggest that the song spoke directly to his relationship with girlfriend Marilyn Rovell, his father Murry, or the Beach Boys themselves.
14 The Beach Boys, *Beach Boys Today!*
15 This would be the Beach Boys final recording before turning to work on *Pet Sounds*.
16 *Brian Wilson Songwriter 1969–1982*.

17 *Brian Wilson Songwriter 1969–1982*.
18 Lambert, 2007, p. 250. The Beach Boys had in fact already produced albums focused on a single theme, the most obvious of which was the 1963 release *Little Deuce Coupe*.
19 The information compiled and presented in this section of the essay draws on several sources including Badman, 2004; Carlin, 2006; Crowley, 2015; Dillon, 2012; Fusilli, 2005; Granata, 2003; Lambert, 2007; Leaf, 1978; Leaf, 1990; Leaf, 1993; Leaf, 2003.
20 Master recording number: 55558, take 21; composer(s): Brian Wilson, Tony Asher, Mike Love; recording venue(s): Gold Star Studios; backing tracks recorded: January 22, 1966; vocals recorded: March 10, 1966 and April 11, 1966; lead vocal(s): Brian Wilson (verses), Mike Love (bridge). Session musicians included Hal Blaine (Drums), Frank Capp (Bells, Tympani, Percussion), Lyle Ritz (String Bass), Carol Kaye (Electric Bass), Jerry Cole and Bill Pittman (Guitars), Barney Kessel and Ray Pohlman (Mandolins), Al de Lory (Piano), Larry Knechtel (Organ), Carl Fortina and Frank Marocco (Accordions), Steve Douglas, Plas Johnson and Jay Maigliori (Saxophones), and Roy Caton (Trumpet).
21 Leaf, *Pet Sounds – Liner Notes*: 9. Master recording number: 55314, take 23; composer(s): Brian Wilson, Tony Asher; recording venue(s): Western Studios; backing tracks recorded: November 1, 1965 and January 24, 1966; vocals recorded: January 24, 1966; lead vocal(s): Brian Wilson. Session musicians included Hal Blaine (Drums), Jerry Williams (Percussion), Julius Wechter (Tympani, Latin Percussion), Lyle Ritz (String Bass), Carol Kaye (Electric Bass), Jerry Cole, Barney Kessel and Billy Strange (Guitars), Al de Lory (Harpsicord), Steve Douglas (Clarinet); Jay Migliori (Bass Clarinet), Bill Green, Jim Horn and Plas Johnson (Saxophones), and Roy Caton (Trumpet).
22 Master recording number: 55591, take 15; composer(s): Brian Wilson, Tony Asher; recording venue(s): Western Studios; backing tracks recorded: February 15, 1966; vocals recorded: February 15, 1966; lead vocal(s): Mike Love (verses), Mike Love and Brian Wilson (choruses); session musicians included Dennis Wilson (Drums), Carl Wilson (Guitar), Brian Wilson (Organ), Lyle Ritz (String Bass), Carol Kaye (Electric Bass), Glen Campbell (12-string Electric Guitar), and Frank Capp (Percussion).
23 Master recording number: 55597, take 1; composer(s): Brian Wilson, Tony Asher; recording venue(s): Western Studios; backing tracks recorded: April 3, 1966; vocals recorded: April 3, 1966; lead vocal(s): Brian Wilson; session musicians included Hal Blaine (Drums), Steve Douglas (Percussion), Frank Capp (Vibraphone, Tympani), Lyle Ritz (String Bass), Carol Kaye (Electric Bass), Glen Campbell, Billy Strange (Guitars), and Al de Lory (Organ).
24 Master recording number: 55865, Take 14; composer(s): Brian Wilson, Mike Love; recording venue(s): Western Studios; backing tracks recorded: March 6, 1966; vocals recorded: March 10, 1966; lead vocal(s): Brian Wilson; session musicians included Jim Gordon (Drums), Gary Coleman (Tympani, Bongos), Carol Kaye (Electric Bass), Ray Pohlman (Guitar), Lyle Ritz (Ukulele), Al de Lory (Piano), Larry Knechtel (Organ), Bill Green, Jim Horn, Jay Migliori (Flutes), and Leonard Hartman (English Horn).
25 Master recording number: 55557, take 18; composer(s): Brian Wilson; recording venue(s): Western Studios; backing tracks recorded: January 18 and 19, 1966; session musicians included Hal Blaine (Drums), Julius Wechter (Tympani, Vibraphone), Lyle Ritz (String Bass), Carol Kaye (Electric Bass), Al Casey, Barney Kessel (Guitars), Barney Kessel and Ray Pohlman (Mandolins), Al de Lory (Piano), Larry Knechtel (Organ), Steve Douglas and Plas Johnson (Tenor Saxophones), Jim Horn and Jay Migliori (Baritone Saxophones), Roy Caton (Trumpet), Arnold Belnick, James Getzoff, William Kurasch, Leonard Malarsky, Jerome Reiser, Ralph Schaeffer, Sid Sharp, Tibor Zelig (Violins), Joseph DiFiore and Harry Hyams (Violas), Justin DiTullio and Joseph Saxon (Cello), Steve Douglas and Jules Jacob (Flutes).
26 For a review of the recorded musical versions of this song, please see Dillon, 2012, pp. 84–85.
27 *Brian Wilson Songwriter 1969–1982*. Master recording number: 53999, take 14; composer(s): arrangement Brian Wilson; recording venue(s): Western Studios; backing

tracks recorded: July 12, 1965; vocals recorded: December 22, 1965; lead vocal(s): Brian Wilson and Mike Love; session musicians included Hal Blaine (Drums), Ron Swallow (Tambourine), Lyle Ritz (String Bass), Carol Kaye (Electric Bass), Al Casey, Jerry Cole and Billy Strange (Guitars), Al de Lory (Organ), Frank Capp (Glockenspiel), Jay Migliori (Clarinet), Steve Douglas, Jim Horn (Flutes), and Jack Nimitz (Baritone Saxophone).

28 Master recording number: 55849, take 20; composer(s): Brian Wilson, Tony Asher; recording venue(s): Western Studios; backing tracks recorded: March 9, 1966; vocals recorded: March 10, 1966 and April 11, 1966; lead vocal(s): Carl Wilson; session musicians included Hal Blaine (Drums), Jim Gordon (Percussion), Lyle Ritz (String Bass), Carol Kaye (Electric Bass), Ray Pohlman (Danelectro Bass), Jerry Cole and Bill Pittman (Guitars), Don Randi (Piano), Larry Knechtel (Organ), Carl Fortina and Frank Marocco (Accordions), Leonard Hartman (Clarinet, Bass Clarinet), Bill Green and Jim Horn (Flutes), Alan Robinson (French Horn), Jay Maigliori (Baritone Saxophone), Leonard Malarsky and Sid Sharp (Violins), Darrel Terwilliger (Viola), and Jesse Erlich (Cello).

29 Master recording number: 55596; composer(s): Brain Wilson, Terry Sachen, Mike Love; recording venue(s): Western Studios; backing tracks recorded: February 9, 1966; vocals recorded: February 9, 1966 and March 1966.; lead vocal(s): Mike Love and Al Jardine (verses), Mike, Love, Al Jardine and Brian Wilson (choruses); session musicians included Hal Blaine (Drums), Julius Wechter (Percussion), Lyle Ritz (String Bass), Ray Pohlman (Electric Bass), Glen Campbell and Barney Kessel (Guitars), Al de Lory (Tack Piano), Larry Knechtel (Organ), Tommy Morgan (Harmonica), Steve Douglas, Jim Horn, Paul Horn and Bobby Klein (Tenor Saxophones), and Jay Migliori (Baritone Saxophone).

30 Master recording number: 55680, take 11 and 20 blended for master; composer(s): Brian Wilson, Tony Asher; recording venue(s): Sunset Sound Studios; backing tracks recorded: March 10, 1966; vocals recorded: March 25, 1966; lead vocal(s): Mike Love; Session Musicians included Nick Martinis (Drums), Frank Capp (Percussion), Terry Melcher (Tambourine), Lyle Ritz (String Bass), Carol Kaye (Electric Bass), Al Casey and Mike Deasy (Guitars), Don Randi (Piano), Larry Knechtel (Organ), Jay Maigliori and Jack Nimitz (Baritone Saxophones), Gail Martin (Trombone), and Ernie Tack (Bass Trombone).

31 According to Tony Asher, "the theremin, a pioneering electronic instrument, was invented by Leon Theremin in Russia in 1920. The instrument was not touched by the player. Its eerie, wailing sound was created by altering the distance of one's hands from the instruments two antennae – one controlling pitch and the other, volume. Jazz musician Paul Tanner and hobbyist Bob Whitsell developed the electro-theremin in that late 1950s." Dillon, 2012, pp. 93–94.

32 Dillon, 2012, p. 92. Master recording number: 55598, take 6; composer(s): Brian Wilson, Tony Asher; recording venue(s): Gold Star Studios; backing tracks recorded: February 14, 1966; vocals recorded: March 10, 1966 and April 13, 1966; lead vocal(s): Brian Wilson; session musicians included Hal Blaine (Drums, Tympani, Bongos), Frank Capp (Tympani, Latin percussion), Chuck Berghofer (String Bass), Ray Pohlman (Electric Bass), Glen Campbell and Barney Kessel (Guitars), Don Randi (Piano), Mike Melvoin (Harpsichord), Paul Tanner (Theremin), Tommy Morgan (Harmonica) Steve Douglas, Plas Johnson and Booby Klein (Tenor Saxophones), and Jay Migliori (Baritone Saxophone).

33 Master recording number: 55848, take 3; composer(s): Brian Wilson; recording venue(s): Western Studios; backing tracks recorded: November 17, 1965; session musicians included Richie Frost (Drums), Richie Frost, Bill Green and Plas Johnson (Percussion), Lyle Ritz (String Bass), Carol Kaye (Electric Bass), Tommy Tedesco (Acoustic Guitar), Jerry Cole and Billy Strange (Electric Guitars), Brian Wilson (Piano), Larry Knechtel (Organ), Carl Fortina and Frank Marocco (Accordions), Bill Green, Jim Horn and Plas Johnson (Tenor Saxophones), Jay Maigliori (Baritone Saxophone), and Roy Caton (Trumpet).

34 Master recording number: 55536, take 17; composer(s): Brian Wilson, Tony Asher; recording Venue(s): Western Studios; backing tracks recorded: January 31, 1966; vocals recorded: January 31, 1966; lead vocal(s): Brian Wilson; session musicians included Hal Blaine (Drums), Frank Capp (Vibraphone), Carol Kaye (Electric Bass), Glen Campbell and Barney Kessel (Guitars), Lyle Ritz (Ukulele), Al de Lory (Harpsichord), Bill Green, Jim Horn, Plas Johnson and Jay Migliori (Flutes), and Steve Douglas (Tenor Saxophone).
35 "Good Vibrations" was completed and released on October 10, 1966 becoming a #1 hit.
36 The Duophonic method essentially attempted, albeit not very successfully, to simulate a true stereo recording.
37 See The Beach Boys, *The Pet Sound Sessions* (Hollywood: Capitol Records, 1997).
38 See *Brian Wilson Presents Pet Sounds Live in London* (New York: Sanctuary Visual Entertainment, 2003).
39 Perhaps the greatest concept album ever conceived, but never released, would be the Beach Boys album that was scheduled to follow *Pet Sounds*; namely, *Smile*. The reasons behind Brian Wilson and The Beach Boys' decision to effectively shelve the *Smile* project are well documented. A detailed account is included in Lambert, 2007, pp. 260–275.

Bibliographic sources

Audio album/compact disc recordings (Listed according to U.S. Release Date).

The Beach Boys (1962) *Surfin' Safari*. Hollywood: Capitol Records, T1808, 1 Oct.
The Beach Boys (1963) *Surfin' USA*. Hollywood: Capitol Records, T1980, 25 Mar.
The Beach Boys (1963) *Surfer Girl*. Hollywood: Capitol Records, T1981, 16 Sept.
The Beach Boys (1963) *Little Deuce Coupe*. Hollywood: Capitol Records, T1998, 3 Oct.
The Beach Boys (1964) *Shut Down Volume 2*. Hollywood: Capitol Records, T2027, 2 Mar.
The Beach Boys (1964) *All Summer Long*. Hollywood: Capitol Records, T2110, 13 Jul.
The Beach Boys (1965) *Beach Boys Today!* Hollywood: Capitol Records, T2269, 8 Mar.
The Beach Boys (1965) *Summer Days (And Summer Nights!!!)*. Hollywood: Capitol Records, T2354, 5 Jul.
The Beach Boys (1966) *Pet Sounds*. Hollywood: Capitol Records, MAS2398, 16 May.
The Beach Boys (1993) *Good Vibrations: Thirty Years of the Beach Boys*. Hollywood: Capitol Records, D 207100.
The Beach Boys (2001) *Hawthorne, CA*. Hollywood: Capitol Records, CDP 531583.
The Beach Boys (2013) *The Beach Boys Live: The 50th Anniversary Tour*. Hollywood: Capitol Records, B0018419-02.
The Beach Boys (2016) *Becoming the Beach Boys: The Complete Hite & Dorinda Morgan Sessions*. Los Angeles: Omnivore Recordings, OVCD-186.
Wilson, Brian (2000) *Brian Wilson Live at the Roxy Theatre*. Los Angeles: Brimel Records.
Wilson, Brian (2016) *Brian Wilson and Friends*. London: BMG.

DVD audio/video recordings

The Beach Boys 50 Live in Concert (2012) San Francisco: Brother Records Inc.
The Beach Boys Good Timin' Live at Knebworth, England 1980 (2003) London: Eagle Rock Entertainment Limited.
The Beach Boys the Lost Concert (1998) Chatsworth, CA: Sabucat Productions.
Brian Wilson on Tour (2003) New York: Sanctuary Records Group.

Brian Wilson Presents Pet Sounds Live in London (2003) New York: Sanctuary Visual Entertainment.
Brian Wilson Songwriter 1962–1969 (2010) Surrey: Chrome Dreams.
Carl Wilson Here and Now (2011) Las Vegas: MFM Productions, Inc.
Pet Sounds – DVD Audio (2003) Hollywood: Capitol Records.
A Tribute to Brian Wilson (2007) London: Eagle Rock Entertainment Limited.

Print sources

Abbott, K. (2001) *The Beach Boys' Pet Sounds – The Greatest Album of the Twentieth Century*. London: Helter Skelter Publishing.
Badman, K. (2004) *The Beach Boys – The Definitive Diary of America's Greatest Band: On Stage and in the Studio*. San Francisco: Backbeat Books.
Carlin, P. A. (2006) *Catch a Wave: The Rise, Fall & Redemption of the Beach Boys' Brian Wilson*. Emmaus, PA: Rodale Inc.
Crowley, K. (2015) *Long Promised Road: Carl Wilson, Soul of the Beach Boys – The Biography*. London: Jawbone Press.
Dillon, M. (2012) *Fifty Sides of the Beach Boys: The Songs That Tell Their Story*. Toronto: ECW Press.
Fawcett, A. (1978) *California Rock, California Sound: The Music of Los Angeles and Southern California*. Los Angeles: Reed Books.
Fusilli, J. (2005) *Pet Sounds (33 1/3)*. New York: The Continuum International Publishing Group Inc.
Granata, C. A. (2003) *Wouldn't It Be Nice: Brain Wilson and the Making of the Beach Boys' Pet Sounds*. Chicago: A Cappella Books.
Harrison, D. (1997) 'After Sundown: The Beach Boys' Experimental Music', in Covach, J. and Boone, G. M. (eds.) *Understanding Rock: Essays in Musical Analysis*. New York: Oxford University Press, pp. 33–57.
Hartman, K. (2012) *The Wrecking Crew*. New York: St. Martin's Press.
Lambert, P. (2007) *Inside the Music of Brian Wilson: The Songs, Sounds, and Influences of the Beach Boys' Founding Genius*. New York: The Continuum International Publishing Group Inc.
Leaf, D. (1978) *The Beach Boys and the California Myth*. New York: Grosset & Dunlap.
Leaf, D. (1990) *Pet Sounds – Liner Notes*. Hollywood, Capitol Records, CDP 7484212.
Leaf, D. (1993) *Good Vibrations: Thirty Years of the Beach Boys – Liner* Notes. Hollywood: Capitol Records.
Leaf, D. (2003) *Pet Sounds – DVD Audio, Liner Notes*. Hollywood: Capitol Records.
May, K. G. (2002) *Golden State, Golden Youth: The California Image in Popular Culture, 1955–1966*. Chapel Hill: The University of North Carolina Press.
Morgan, J. (2015) *The Beach Boys America's Band*. New York: Sterling.
Murphy, J. B. (2015) *Becoming the Beach Boys, 1961–1963*. Jefferson, NC: McFarland & Company Inc. Publishers.
O'Regan, J. (2016) 'When I Grow Up: The Beach Boys' Early Music', in Lambert, P. (ed.) *Brian Wilson and the Beach Boys in Critical Perspective*. Ann Arbor: University of Michigan Press, pp. 137–167.
Rusten, I. and Stebbins, J. (2013) *The Beach Boys in Concert: The Ultimate History of America's Band on Tour and Onstage*. Milwaukee: Backbeat Books.
Schinder, S. (2008) 'The Beach Boys', in Schinder, S. and Schwartz, A. (eds.) *Icons of Rock: An Encyclopedia of the Legends Who Changed Music Forever*, Volume 1. Westport, CT: Greenwood Press, pp. 101–129.

Simmons, S. (2003) *Brian Wilson Presents Pet Sounds Live in London – Liner Notes*. New York: Sanctuary Visual Entertainment.

Starr, K. (2009) *Golden Dreams: California in an Age of Abundance, 1950–1963*. New York: Oxford University Press.

Starr, L. and Waterman, C. (2014) *American Popular Music: From Minstrelsy to MP3*. New York: Oxford University Press.

Stebbins, J. (2011) *The Beach Boys FAQ*. Milwaukee, WI: Backbeat Books.

Tuttle, P., Samson, V., Hutchison, S., Kirkpatrick, R. and Goggans, J. (2004) 'Music', in Goggans, J. and DiFranco, A. (eds.) *The Pacific Region: The Greenwood Encyclopedia of American Regional Cultures*. Westport, CT: Greenwood Press, p. 358.

Wilson, B. and Gold, T. (1991) *Wouldn't It Be Nice – My Own Story*. New York: HarperCollins Publishers.

Wilson, B. and Greenman, B. (2016) *I am Brian Wilson*. Boston: Da Capo Press.

5 A Western skyline I swear I can see

Affective critical rurality expressed through contemporary Americana music

Keith Halfacree

Introduction: representing 'authentic' rural and small-town America

One only needs to turn on the computer, radio or television to confirm how Donald Trump's 2016 election as 45th U.S. President continues to stimulate huge international interest. One thing his election achieved immediately was to draw more than usual political attention to less metropolitan parts of the U.S. Journalists were drawn to the 'provocative storyline' (Monnat and Brown, 2017, p. 227) of 'rural' voters being one of the core groups – Trump received 63% of the 'rural' vote – that had made a highly unlikely electoral victory possible. This is reinforced by the electoral geography of Trump's victory, whereby an established overall pattern of more rural, more Republican was sharply restated (Scala and Johnson, 2017).

The key **rural** message from the vote, however, was arguably less pro-Trump than anti- a perceived status quo that had allowed 'localized economic distress' (Monnat and Brown, 2017, p. 229) to develop and fester. It expressed a backlash against being ignored and marginalised by a national polity overwhelmingly associated with the values, priorities and peoples of the major cities (Gusterson, 2017). In short, populist self-presentation as an anti-establishment figure – of course, readily disputable – projected Trump as a subversive iconoclastic figurehead for these neglected communities (*ibid.*). Through 'position[ing] himself as an outsider to the political arena and . . . defender of "common men"' (Lamont *et al.*, 2017, p. S166), he 'brought the politically marginalized white working class back to the voting booth by cultivating differences' (*ibid.*: S173), not least across rural America.

Looking beyond the Trump vote, however, there are other ways that the forgotten people of the rural and small-town U.S. challenge, or at least come to some degree of reconciliation with, their marginalisation. They should not, in other words, be dismissed as a 'passive' population or summarily dismissed 'as backward, uneducated, and "redneck"' (Sherman, 2009, p. 4). Consequently, besides mapping their challenges 'on the ground', such as the Trump vote, their voices can be heard through various cultural expressions. This is the focus of this chapter. It explores how voices from what I have termed 'abandoned rural America'

(Halfacree, 2018) are expressed through a musician rooted in the U.S. West. As Scoones *et al.* (2018, p. 12) stated: '[E]mancipatory research of the rural ... [must be] attentive to hinterlands, margins and frontiers'.

The chapter continues by more fully noting the increasingly marginalised everyday lives of many rural and small-town U.S. residents, especially drawing out experiences from the West. This situation is contrasted with that region's strongly positive symbolic image in the Frontier myth, which remains powerfully embedded within mainstream U.S. culture. Country Music as a cultural form engaging both these perspectives is introduced, drawing out its Americana offshoot as its most 'authentic' expression. After presenting the concepts 'reading for difference' and 'affective critical regionality', these are deployed in analysis of Americana musician Willy Vlautin's songs. This draws out four key themes related to abandoned rural America. All portray the hardship of lives, but ultimately in ways that contrast with mainstream accounts. The fourth theme, moreover, articulates how everyday life tactics can reap existential rewards from brief disruptions to normal everyday life. In conclusion, it is argued that 'freedom' for many within abandoned rural America today is to appreciate modest rewards from an 'affective critical rurality'.

Expressing economic marginalisation in rural and small-town U.S.

Distress...

Whilst the literature remains **relatively** sparse when considering the huge areas, large numbers of people and great diversity of situations it needs to cover, there are now many informative studies outlining the ongoing economic marginalisation of many people in rural and small-town U.S. and the consequences of their abandonment. Evidence clearly refutes commonplace assumption of U.S. poverty 'as an urban minority problem' (Sherman, 2006, p. 891). Yet, it remains true that corresponding (and, of course, invaluable) research and policy initiatives focused on such urban populations still make the rural poor 'largely invisible to some' researchers and policymakers (Burton *et al.*, 2013, p. 1171).

Rural research, on the one hand, has continued to focus on widely acknowledged long-standing impoverished areas, such as Appalachia, parts of New England and the Mississippi Delta. Such scholarship's continued salience is exemplified through publication of a second edition of Cynthia Duncan's pioneering and hard-hitting qualitative account of rural poverty, *World's Apart* (Duncan, 2015). On the other hand, scholars have demonstrated an ongoing geographical and social expansion of rural poverty, pioneered by Janet Fitchen's recognition of the legacy of the 1980s farm crisis in rural New York state and beyond, *Endangered Spaces* (Fitchen, 1991). As Elliott (1994, p. 129) ably summarises, Fitchen presented a

> rural America that more closely resembles ... images of metropolitan USA than ... mythic notions of the rural countryside ... [through] depicting

growing poverty, a greying population, households increasing in number faster than the number of people, out-migration of many of the most able, and in-migration of either those unable to make it elsewhere or "weekend people" (p. 101) who raise housing prices and the need for local services while contributing little to the local economy or community.

Both the ingrained and growing condition of rural poverty has been placed within a 'new rural order' (Burton *et al.*, 2013, p. 1129) shaped by economic restructuring and low-income in-migration. As spatial shifts in rural poverty take in 'new' places, sustained attention is being accorded to the full range of rural and small-town locations where 'economic distress has been building, life expectancy has been declining . . . and social conditions have been breaking down for decades' (Monnat and Brown, 2017, p. 229). Distress, exacerbated by the 2007–09 Recession (Burton *et al.*, 2013; Carr and Kafalas, 2009; Ulrich-Schad and Duncan, 2018), is mapped out by direct expressions of poverty and lack of work but also by problem-reinforcing situations of declining health, rising alcohol and drug dependency, poor schooling, out-migration 'brain drain' of the most qualified, growing crime rates and rising 'deaths of despair' (Scoones *et al.*, 2018, p. 6).

Indeed, rural distress is never solely economic but also a matter of culture and morality, as powerfully detailed by Jennifer Sherman (2006, 2009, 2018, etc.). Drawing on in-depth research in the pseudonymous Golden Valley community in the northern California woods, and emphasising that this is not a special case, Sherman has articulated how community impacts of extreme economic stresses are not solely material. Specifically, **moral capital** within everyday life counts for much in Golden Valley, not least with opportunities for acquiring economic capital highly limited. To gain and retain such capital, the poor are 'compel[led] . . . to act in mainstream ways' (Sherman, 2006, p. 907), with strong 'social pressure . . . to be culturally acceptable according to the existing local standards' (Sherman, 2006, p. 893). Moreover, with few well-paid or secure jobs, 'morality aids in the creation of symbolic boundaries [hierarchies] between groups of people for whom there are few other forms of distinction available' (Sherman, 2009, p. 4).

Sherman's moral capital is particularly seen as rooted in regular and reliable undertaking of hard manual work, especially by men (Sherman, 2009; Ulrich-Schad and Duncan, 2018), reconnecting cultural and economic (as her Bourdieusian analysis would anticipate). However, opportunities for even the most 'moral' to attain 'distinction' through acquiring such capital are increasingly narrowing. On the one hand, the rural economy is shifting from 'good' manufacturing and primary industry jobs towards those in the service sector. The latter are often not only both part-time and low wage, but also largely taken-up by poor in-migrants (Ulrich-Schad and Duncan, 2018). On the other hand, in-migration of wealthier U.S. citizens, particularly to high natural amenity areas (*ibid.*), bolsters a social divide. Although reinforcing the importance of moral capital for the poor – lacking newcomers' economic capital – it paradoxically ultimately reinforces the status of economic capital through sharply dividing the population (Sherman, 2018). In sum, the U.S. rural poor – estimated a few years ago as 7.9 million people

(Burton *et al.*, 2013) – are being comprehensively 'left behind in the new economy' (*ibid.*, p. 1132).

Yet, geographical resilience . . .

A moral capital emphasis helps explain how 'ties to place, and deeply value[d] family and community' (Ulrich-Schad and Duncan, 2018, p. 60) keep many of the poorest in abandoned rural America rather than prompting their movement away. A vital further cultural reinforcement, however, is adherence to elements of the American Dream (Sherman, 2009). Nowhere is this association stronger than in the West, the most mythologised U.S. region (Slatta, 2010). This region is the geographical nexus of the powerful Frontier myth and forms the geographical backdrop for the rest of this chapter.

Frederick Jackson Turner's 1893 *Frontier Thesis* sought to associate U.S. democracy, egalitarianism and, building on this, the very cultural essence of the United States itself with pioneers' struggles to overcome and push forward its geographical frontier. He 'made the frontier experience **the** central force in creating American identity and values' (Slatta, 2010, p. 83), sowing the seeds for a 'creation story for America' (Campbell and Kean, 2006, p. 138) overall. From the early settled eastern U.S., the West was positioned as a primitive wilderness but, nonetheless, the arena in which the men (*sic.*) who progressively 'conquered' and 'civilised' it through settlement articulated core values of freedom, independence, individualism, competition and self-sufficiency that became engrained across U.S. culture.

The Frontier myth has, of course, been subject to widespread deconstruction, not least by drawing out the geographical un-exceptionality of many of the lives lived and the significance of social cooperation at least as much as individualistic competition (Slatta, 2010). It is, nonetheless, a place myth that strongly persists (Campbell and Kean, 2006), not least on account of its strong presence within popular culture, from Westerns, through photographs and novels, to a powerful discursive ideology articulated by politicians such as Trump. The Frontier, in short, is a 'concept . . . here to stay' (Slatta, 2010, p. 88) and it is foolish to try to 'bury or ignore' (*ibid.*: 89) it. Certainly, it is intimately entangled with the striving for moral capital in the context of the rural poor's efforts to live with a degree of 'respect'. Combined with political conservatives' stress on 'family values', Frontier 'individualism, self-sufficiency, and work in the form of manual labor as both moral and masculine' (Sherman, 2009, p. 10) are very much alive across abandoned rural America (and beyond). They are certainly articulated strongly within Country Music.

Americana within Country Music

'Country Music' is a broad church. As Wikipedia (2018) aptly notes, it covers

> many styles and subgenres . . . [reflecting] origins . . . [in] the folk music of working class Americans, who blended popular songs, Irish and Celtic fiddle

tunes, traditional English ballads, cowboy songs, and various musical traditions from European immigrants.

Indisputable is its immense popularity and huge market in the U.S. and beyond, with annual U.S. album sales alone regularly exceeding 40 million (Statista, 2019). However, whilst Country Music frequently expresses rural and small-town economic decline, depressed living conditions and resentment towards perceived elites (Bleakley, 2018), how it does so is also far from consistent.

Country's Americana sub-genre – itself diverse and eclectic, as reflected in alternative names: Alt.country, Alternative Country, No Depression (Goodman, 1999; Holt, 2007; Peterson and Beal, 2001) – emphasises 'a rhetoric of taste, ties to country tradition, and the cultivation of a contemporary, discerning community of liberal-minded fans' (Ching and Fox, 2008, pp. 3–4). These features gel into a dominant trope of 'cultural authenticity' (Kirby, 2006) and a mission to reclaim Country Music as the voice of the American working-class (Cooper, 2012). Whilst Americana is certainly not exclusively rural and small-town, it is strongly focused on and rooted in local places and geographically specific experiences. Unsurprisingly, therefore, it has much to say about lives across abandoned rural America. This will shortly be illustrated through Willy Vlautin's songs, but the chapter first takes a short intermission to introduce, first, a concept that will underpin analysis of these songs and, second, a concept to express a sense of the rural and small-town West different from that articulated by the mainstream.

Conceptual intermission: 'reading for difference' for an 'affective critical regionality'

'Reading for difference' is an analytical strategy fundamentally opposed to the perceived and perhaps inevitable (over)simplification inherent to generalisation. It underpinned the long-term project of Katherine Gibson and the late Julie Graham – as J.K. Gibson-Graham (Community Economies, 2018) – that developed their feminist-inspired critique of established political economy's understanding of economies as **singularly** capitalist. Gibson-Graham emphasised the diversity found within **all** economies and the value of recognising these multiple economic life forms. From their critical anti-capitalist stance, they sought to promote some of these alternatives, not least in communities failed by the dominant capitalist model and language (Community Economies, 2018). More generally, everyday life is recognised as not only capitalist, with some of its more-than-capitalist elements again meriting greater recognition.

One means to challenge the totalising and encompassing character of the capitalist meta-narrative but also, by extension, other equally singular representations is through '[r]eading for difference rather than dominance' (Gibson-Graham, 2006, pp. xxxi–xxxii). Instead of reiterating and consequently reinforcing – often unintentionally – ubiquitous dominant representations, the focus is on uncovering and delineating 'what is possible but obscured from view' (*ibid.*: xxxi). Through this, 'future possibilities become more viable by virtue of already being seen to exist' (*ibid.*).

There are many situations whereby an already-existing but neglected 'minor language' (Deleuze and Guattari, 1987, p. 105; also Campbell, 2016) can become better acknowledged and acquire greater potential to challenge 'major language' dominance. Within even the most mundane daily life mainstream, subversive currents frequently lurk. Such currents have been teased out by Neil Campbell (2016) – a leading scholar of cultural expressions of the U.S. West – through his 'affective critical regionality' concept.

To better capture the inherent diversity of life within the West, Campbell sought a term expressing a humanism rooted in 'contingency, precarity, and vulnerability' (Campbell, 2016, p. 4) and set explicitly within the specificity of local place, altogether refusing 'uncontested and absolute vision' (Campbell, 2016, p. 13). Whilst not denying the powerful role played by the Frontier myth, Campbell argued – akin to Gibson-Graham – for it not to be seen as the only game in town. Instead of **any** settled 'regionalism' expressed in such a major language, affective critical regionality 'conjoins the region with the felt, dynamic, turbulent, responsive and heterogeneous sense of "-ity" [rather than with] the taken-for-granted, static, settled, nostalgic and uniform "-ism"' (Halfacree, 2018, p. 4). Regionality is fundamentally felt, lived at least as much as stemming from representation: it is atmospherically affective (Anderson, 2009). Further, Campbell's concept recognizes the radical disruptive power of minority currents through the addition of 'critical' to the term.

Reading for difference to express affective critical regionality requires looking for expressions of particular spaces and lives lived that emphasise not only their diversity but also their grounded deviation from the status quo. The chapter will now begin to do this through reading for difference in stories told in musician Willy Vlautin's songs. Extending the practice's usual single minor language excavation, this involves two iterations. First, the songs will be argued as expressing lives across the West that challenge any all-round naive celebration of this landscape. This affiliates them with Country Music's quasi-Frontier narrative of living hard lives in hard places, albeit typically romantically articulated in a manner Vlautin and Americana generally avoid. Second, Vlautin's own representations are themselves then argued to contain their own minor language, involving the place-rooted temporary seizure of hope within bleak everyday lives.

Reading authentic Western lives: illustrations from Willy Vlautin

Introduction: Willy Vlautin and his music

Born in 1967, Willy Vlautin is now widely recognised as an award-winning novelist. Two of his five books to date have also been turned into movies, notably the Andrew Haigh-directed *Lean on Pete* (2018). Focus here, though, will be on his songs, helping to complement the attention his novels are attracting in the academic literature (for example, Campbell, 2018).

A Western skyline I swear I can see 67

Vlautin was raised in Reno, Nevada, by a single mother in precarious employment (Gibney, 2015). Subsequent experiences reinforced his positional suitability for writing the songs noted below. Although Reno was relatively prosperous, Vlautin grew up around many who had fallen on hard times. Gambling, a central Reno feature, appealed strongly, but so did life lived through music, books and films (O'Hagan, 2016, np). After holding various jobs giving further insights into working-class lives, Vlautin relocated to Portland, Oregon, where in 1994 he co-founded the Americana band Richmond Fontaine (Berhorst, 2002). Table 5.1 lists their principal releases, plus those of Vlautin's current band, The Delines.

With the notable exception of those set in Reno, Vlautin's songs are mostly located 'in the great swarming emptiness of an unlit America' (Jones, 2009, np), the inbetween places that epitomise abandoned rural America. More specifically, they are set predominantly within and around the *Thirteen Cities* of Richmond Fontaine's 2007 album. Importantly for this chapter's rural focus, whilst seven of the cities' populations exceed 200,000 – Phoenix, at over 1.5 million, by far the largest – four (Bullhead City, Arizona; Walla Walla, Washington; Laramie, Wyoming; tiny Mojave, California) have populations of no more than 40,000.

Songs typically express 'semi-autobiographical blue collar tales of lives on the edge' (Clarkson, 2005, p. 3), and characters 'worried and concerned and scared' (Vlautin, in McGrath, 2015, np). Vlautin also articulates critical social messages, expressed in themes such as the struggle for workers' rights and scepticism towards U.S. foreign policy and wars where 'working class guys get killed, and rich people get richer' (Vlautin, in Gibney, 2015, np). He portrays working people betrayed by the American dream but struggling and somehow getting through in often very individual and inward-looking ways rather than by expressing a reactive and aggressive striking back. Vlautin's liberal humanism acknowledges how

Table 5.1 Albums of Richmond Fontaine and The Delines

Title	Year
Safety	1996
Miles From	1997
Lost Son	1999
Winnemucca	2002
Post to Wire	2004
The Fitzgerald	2005
Thirteen Cities	2007a
$87 and a Guilty Conscience that Gets Worse the Longer I Go E.P.	2007b
We Used to Think the Freeway Sounded Like a River	2009
The High Country	2011
You Can't Go Back If There's Nothing to Go Back To	2016
Don't Skip Out on Me	2018
Colfax by The Delines	2014
Scenic Sessions by The Delines	2015

Note: individual songs in text are dated by the album they are found on.

even in the worst contexts '[y]ou meet people every day who are kind and decent' (Vlautin, in McGrath, 2015, np).

Freeman (2014, p. 2) insightfully summarises Vlautin as a 'champion of the plight of America's underclass', with his songs teasing out 'the tiniest shred of beauty from the most depressing situation'. Developing this, four themes capturing lives across the rural West have been identified within Vlautin's compositions (Halfacree, 2018). Two paint a grim picture of everyday hardships in tune with contemporary academic accounts but strongly counter to Frontier myth positivity. A third at first seemingly follows the myth through emphasis on mobile struggle but ultimately diverges through suggesting no positive resolution, even for 'deserving' individuals. A fourth theme reads the first three themes for differences and is the most differentiated from the Frontier myth. It expresses the rewards of disengagement through temporary uncertain spatio-temporal displacement from quotidian hardship.

Travelling 'a black road': internal and external victimhood

The first two themes portray a wide variety of the human victims frequently presented across abandoned rural America. A dismal array of alcoholics, drug abusers, battered wives and children, lonely and isolated individuals, the sick, in-debt gamblers, deserted partners and the mentally unstable all travel 'a black road' ('Black Road', 2005) through a landscape of despair. Victimhood comes both internally and externally, via localised inward-looking forces and/or via more global and outward-originating threats. Internal victimhood frequently speaks through excessively parochial lives, whereby lived places at the time depicted are opportunity-free and emotionally barren. Living in such places too long risks one becoming 'obliterat[ed] by time' ('Settle', 1996), destined to drift as a 'ghost' through life ('A Ghost I Became', 2007a; 'I'm Just A Ghost', 2014), 'belted and constricted' ('Evergreen Power Line', 1997) by repressive parochiality. Protagonists are never able to become more than voyeurs of the American dream. As Sherman recognised elsewhere, such items as 'parked cars and lawns and other people's homes' ('St. Ides', 'Parked Cars' and 'Other People's Homes', 2007a) come to 'symbolize . . . opportunities lost to them' (Sherman, 2018, p. 191). Such isolation is frequently reinforced by relationship breakdown, alcoholic ill health ('Two Alone', 2009; 'A Night in the City', 2016), sexual promiscuity ('Willamette', 2004; 'The Boyfriends', 2009), even madness ('Hallway', 2004) as one 'disappear[s] into heartache' ('Disappeared', 2005). Present despair is also often given a strong temporal dimension, typically sharpened through depicting a hometown return where the broke and homeless protagonist finds nothing supportive any more ('Black Road', 2005; 'Got Off the Bus', 2016). Thus one readily becomes sympathetic with the despair 'Wilson Dunlap' (2007b) has over those for whom 'it feels like nothing's ever going to turn out right'.

External victimhood has Vlautin depicting dwelt places lacking key individuals and opportunities, pushing both people and place to fall apart. First, work demands call people away and disrupt emplaced lives: a truck driver 'never calls home' but engages truckstop prostitutes whilst his wife 'falls to her knees . . . drinks vodka, takes Valium and watches TV' ('White Line Fever', 1996); being 'never . . . so uncertain or scared or alone' after moving to work in the city ('Montgomery Park', 2003); or friends displaced by involvement in the drugs trade ('Lost Son', 1999). Second, emotional absences disrupt everyday dwelling: missing lovers, where 'the world unravels without you' ('Wichita Ain't So Far Away', 2014), consequent 'isolation . . . my biggest fear' ('Northline', 2002); or impacts of other missing family, from a Merchant Marine brother ('Willamette', 2004), to sitting with a U.S. Army deserter cousin mourning a dead brother ('Exit 194B', 2005), to seeking a vanished father ('Polaroid', 2004), to grieving even the abducted ('Trembling Leaves', 1997) or a disappeared neo-Nazi convert ('The Disappearance of Ray Norton', 2007a). Third, dwelt places are pulled apart by external forces, especially economic, the precarious 'The Warehouse Life' (2005) indicative of general economic insecurity. Even those who have returned are often scarred, like exservices speed-addict Angus King not having 'left his house . . . since 2003' ('The Chainsaw Sea', 2011). Fourth, diverse external threats also come from potential arrest for 'growing weed' to stave-off bankruptcy ('43', 2009), having a partner pining for her son 1,000 miles away ('The Longer You Wait', 2004), even urban sprawl ending the (rural) West ('The Water Wars', 2007b).

Both victimhood themes might at first seem potentially in-line with the dominant language of an abandoned rural America widely articulated by mainstream Country Music and by both Trump and interpreters of his rural success. However, there are notable differences. The internal victimhood theme clearly portrays hardship less romantically than mainstream Country. Nonetheless, neither does it adopt the harshly judgmental moral line of the Frontier myth, whereby, for example, alcoholism becomes 'a sign of weak moral character and poor work ethics' (Sherman, 2009, p. 22). The external victimhood theme is presented in much less anti-global and/or anti-immigrant terms than by the political right (Bleakley, 2018; Lamont *et al.*, 2017), whilst still relating to lived consequences of ongoing economic restructuring, in particular.

Overall, in spite of their differences from the mainstream, Vlautin's Western stories nonetheless still seemingly speak with little hope for their characters. However, these characters are not passive victims but people who stay 'in the fight . . . get up each morning and try and get to a better place' (Vlautin, in Freeman, 2014, p. 6). Some do this through embracing mobility, a third theme in the songs.

'It ain't always wrong to give up and run': mobilising to escape

Settlement and developing deep-rooted connection to place form a conclusive element of the Frontier myth – farmer replaces pioneer adventurer – but the core idea of a shifting Frontier also emphasizes the importance of personal mobility within the myth (Campbell and Kean, 2006). With mobility now widely recognised as a key trope of our times – an era of mobilities (Barcus and Halfacree, 2017, Chapter 5) – it is unsurprising that it is expressed by Vlautin as a key means for the marginalised to fight back. 'White Line Fever' (1996), 'Under Florescent Lights'

(1997) and 'Willamette' (2004) all contrast geographically constrained lives with other family members' mobility. 'However, the ultimate fate of those who become mobile goes on to qualify the ultimate success and value of embracing mobilities.' In sum, potential for and actual flight offer renewed hope but little certainty in Vlautin's songs; seeking the 'Western Skyline' (2002) can be cruelly deceptive.

Flight pervades *The High Country* LP. However, whilst 'The Mechanic's Life' (2011) concludes '[i]t ain't always wrong to give up and run', when he flees with his lover it results in murder. More generally in Vlautin's songs, an atmosphere of irresolution pervades the runaway escaping everyday brutality to join his aunt on her ranch ('Laramie, Wyoming', 2005), the yearning for troubled friends to be 'Always on the Ride' (2004), a sister's 4am flit after an 'emergency' ('Casino Lights', 2005), and a wife's dream of fleeing before her husband's return from the Gulf ('The Oil Rigs At Night', 2014). Frequently, initial hope also rapidly becomes compromised: an abused woman's placeless 'foreign' life requires 'nights in the bathtub just to calm her nerves' ('Don't Look and it Won't Hurt', 2005); fleeing a broken home, a boy wrecks the car he drives for an 'old American lady', killing her ('Fifteen Year Old Kid in Nogales, Mexico', 1999); or a desperate woman makes a final call home, 'redemption failed and unattainable' ('Watsonville Waltz', 1996).

Even in an era of mobilities, it therefore seems, taking flight is not a possibility for many and stays a dream. The narrator in 'Give Me Time' (1997) cannot go further than 20 miles beyond Winnemucca before memory pulls him back to the 'bright lights' (*sic*.), whilst the protagonist in 'State Line' (2014) frustratingly 'still can't escape . . . home' psychologically, even though 'no place do I find home'. Elsewhere, whilst a Native American similarly feels nowhere as home, 'every time I try to leave I always end up back at this place' ('Song for James Welch', 2007b).

Besides out-migration, actual or suppressed, characters also deploy other expressions of mobility to try to come to terms with their lives. The bus, in particular, provides a mobile space-time for Vlautin to reflect on life. Again, this is qualified, subsequent anxiety leading to self-harm ('Ft. Lewis', 1996) or to disembarking in just a T-shirt in a snowstorm ('Five Degrees Below Zero', 2002)! Even safe arrival can result in discovering 'nothing to go back to' ('I Got Off the Bus', 2016). Mobility generally resolves as little as flight.

In sum, whilst Vlautin's foregrounding of mobilities may at first sight express the Frontier myth's 'mobility and regeneration' (Río Raigadas, 2016, p. 45) association in a contemporary guise, this is challenged by regeneration almost never being realised. Indeed, 'sometimes . . . if you leave . . . you end up . . . worse [than] if you give up' ('Five Degrees Below Zero', 2002). All one gets from 'a never ending haul of movement where nothing appeared except 13 cities in 7 years' ('Four Hours Out', 1999) is a 'motel life' (Vlautin, 2006) – not 'much of a life . . . a motel ain't much of a home' ('Westward Ho', 2007a) – and the loss of jobs, friends and savings ('He Told Her the City was Killing Him', 2014). In short, actively seeking the mythical 'Western Skyline' (2002) does **not** take Vlautin's protagonists somewhere they 'will be and . . . be set free'. Instead, it might often be wiser to 'give up'.

Heterotopic embraces within 'giving up': an affective critical rurality

Early in his life, Vlautin found 'redemption . . . in . . . music' (Brannan, 1996, np) as a counter to often challenging life experiences. This sense of an ability to escape into and seek personal value from something familiar and relatively mundane is a minor story running through many of the lives told by his songs. Specifically, characters find existential revitalisation, even just the will and ability to go on, from places and times bracketed-out and yet still within their 'normal' everyday lives. Such pauses within their 'giving up', whilst ultimately often at considerable cost financially or in health terms, liberate an 'affective atmosphere' (Anderson, 2009) of content, comfort and consolation. These rewarding glitches within an otherwise seemingly 'endless' depressed dynamic are the opposite of within-*zeitgeist* turns to mobilities. Rather than move when one 'feel[s] so through' ('Through', 2004) in 'a lost and spinning world full of unquestionable brutality' ('Hope and Repair', 1999; 'Lost in this World', 2007a), a response can be to restate local embeddedness and reassert positive existential values in a familiar place.

First, home can express such a recuperative space. Thus, the distant big city voice in 'You Can Move Back Here' (2009) is 'shaky and weird' and begged to return to where 'you don't have to be anything' but 'have the [distanced] Western sky and me on your side'. Elsewhere, acknowledging that '[w]ars will never stop nor destruction and pain', when a partner comes home 'I'll just drown in your arms' ('Flight 31', 2014). Or, after a night out, home 'lights . . . covered and dim . . . [provide] nothing but a gentle ease . . . you and me and our whole place, we're okay' ('Making it Back', 2005).

Second, within 'Four Walls' (2007a) a modest single room can even provide the recuperative space: a bedroom, to '[s]hut the curtains, put a blanket over them, we don't need to see that today' ('Out of State', 2002) or to drink gin with a partner after 'Calling In' (2014) sick to 'stay in bed and watch the day fade'; a motel room sanctuary for a possibly dying battered wife ('The Janitor', 2005); a kitchen ('Making it Back', 2005); or the apartment where the narrator wants his brother to stay with him ('Under Florescent Lights', 1997). Simply, a room is 'a place for me and you . . . We won't bother anyone and everyone will just let us be' ('I Can See a Room', 2011).

Third, other spaces also provide precarious existential recuperation when one 'need[s] some time to drop below that line' ('Winner's Casino', 2002). These include: bars – unsurprisingly, given their prominence in Country Music overall – from 'Let's Hit One More Place' (2016) to rekindle togetherness, to 'Calm' (1997) seeking 'bars where it must be dark and the music must be slow'; a trespassed deserted house ('We Used to Think the Freeway Sounded Like a River', 2009), with a swimming pool ('El Cortez', 1999); late-night walking ('Calm', 1997) or river swimming ('Gold Dreaming', 1996); a phone booth 'Somewhere Near' (2002), where embracing a frozen partner reasserts how 'I knew I could feel good'. More dramatically, one can embark on a '[t]hree day vacation' gambling in nearby Winnemucca, where '[w]hile you're sitting next to me laughing . . . I'm barely losing' ('Barely Losing', 2004).

All told, at least for a while, and certainly not for long-term sustainability, the best immediate hope Vlautin suggests for many across abandoned rural America is neither to attempt to (re)assert mobile and often aggressive Frontier culture nor to strike out at 'external' economic and political interests. It is to 'rediscover' within the relatively local lived elements that can provide temporary modest existential reward. It is to reengage emotionally with place by climbing aboard 'life rafts' (Halfacree, 2018) that provide heterotopic 'counter-sites' (Foucault, 1986). Still part of the protagonist's everyday reality (such as the home), these become spaces 'of deferral . . . where ideas and practices that represent the good life can come into being' (Hetherington, 1997, p. ix). Associating with Campbell's (2016) affective critical regionality, Vlautin's cautious recognition and celebration of such 'heterotopic vacationing' speaks, in sum, of potentially finding in the West an affective critical **rurality** to keep the spirit alive.

Conclusion: 'freedom' *within* the Western Skyline?

In today's 'authentic America', many rural and small-town people are living highly impoverished lives. Neglected rural America's geography now extends well beyond previously noted locations, into even the culturally celebrated heartlands of the West. It is a population, however, that still has a voice and still retains agency as expressive of its humanity. This voice comes through in many ways, including the 2016 Trump presidential vote and right-wing populist discourse. However, this chapter has exemplified its different expression within popular culture through the songs of long-term Western resident, Willy Vlautin. In line with Country Music, especially its Americana sub-genre, Vlautin does not shy away from expressing hardship, both internally and externally brought about, but tells it differently from either right-wing populism or Frontier myth essentialism. He also presents any *zeitgeist* turn to various forms of mobility in response to these conditions less as celebrating rugged individualist freedom but more as frequently misguided and ineffective. Instead, reading for difference within Vlautin's songs pulls out a subdued and uncertain voice that states how the best 'freedom' many can hope to attain, at least in the immediate present, is through modest existential celebration of small, emotionally resonant places and experiences found within an affective critical rurality. Hope, in short, lies more in 'hinterlands, margins and frontiers' (Scoones *et al.*'s, 2018) than in seeking out any renewed mythical Western skyline to 'make America great again'.

Bibliography

Anderson, B. (2009) 'Affective Atmospheres', *Emotion, Space and Society*, 2, pp. 77–81.

Barcus, H. and Halfacree, K. (2017) *An Introduction to Population Geographies: Lives Across Space*. London: Routledge.

Berhorst, K. (2002) 'Interview: Richmond Fontaine', *In Music We Trust*, p. 48. Available at: www.inmusicwetrust.com/articles/48h06.html [Accessed 17 Nov.].

Bleakley, P. (2018) 'Situationism and the Recuperation of an Ideology in an Era of Trump, Fake News and Post-Truth Politics', *Capital and Class* 42(3): 419–34.

Brannan, M. (1996) 'Richmond Fontaine – Safety', *No Depression*, 30 Jun. Available at: http://nodepression.com/album-review/richmond-fontaine-safety [Accessed 17 Nov.].

Burton, L., Lichter, D., Baker, R. and Eason, J. (2013) 'Inequality, Family Processes, and Health in the "New" Rural America', *American Behavioral Scientist*, 57(8), pp. 1128–1151.

Campbell, N. (2016) *Affective Critical Regionality*. London: Rowman and Littlefield.

Campbell, N. (ed.) (2018) *Under the Western Sky: Essays on the Fiction and Music of Willy Vlautin*. Reno: University of Nevada Press.

Campbell, N. and Kean, A. (2006) *American Cultural Studies*, 2nd edition. London: Routledge.

Carr, P. and Kafalas, M. (2009) *Hollowing Out the Middle: The Rural Brain Drain and What It Means for America*. Boston: Beacon Press.

Ching, B. and Fox, P. (2008) 'Introduction: The Importance of Being Ironic – Toward a Theory and Critique of Alt.Country Music', in Fox, P. and Ching, B. (eds.) *Old Roots, New Routes: The Cultural Politics of Alt.Country Music*. Ann Arbor: University of Michigan Press, pp. 1–27.

Clarkson, J. (2005) 'Interview. Richmond Fontaine', *Penny Black Music*, 20 Aug. Available at: www.pennyblackmusic.co.uk/magsitepages/Article/3685/Richmond-Fontaine-Interview [Accessed 17 Nov.].

Community Economies (2018) 'Homepage'. Available at: www.communityeconomies.org/ [Accessed 18 Dec.].

Cooper, T. (2012) ' "Sometimes I Live in the Country, Sometimes I Live in the Town", Discourses of Authenticity, Cultural Capital and the Rural/Urban Dichotomy in Alternative Country Music', MA thesis, Victoria University of Wellington, New Zealand. Available at: http://researcharchive.vuw.ac.nz/xmlui/handle/10063/2378 [Accessed 17 Nov.].

Deleuze, G. and Guattari, F. (1987) *A Thousand Plateaus*. Minneapolis: University of Minnesota Press.

Duncan, C. (2015) *Worlds Apart: Poverty and Politics in Rural America*, 2nd edition. New Haven, CT: Yale University Press.

Elliott, J. (1994) 'Review of Endangered Spaces, Enduring Places', *Journal of Research in Rural Education*, 10(2), pp. 129–130.

Fitchen, J. (1991) *Endangered Spaces, Enduring Places: Change, Identity, and Survival in Rural America*. Boulder: Westview Press.

Foucault, M. (1986/1967) 'Of Other Spaces', *Diacritics*, 16, pp. 22–27.

Freeman, J. (2014) 'A Sliver of Hope: An Interview with Willy Vlautin', *The Quietus*, 14 Apr. Available at: http://thequietus.com/articles/14934-willy-vlautin-interview-the-delines [Accessed 17 Nov.].

Gibney, C. (2015) 'Willy Vlautin: "More the Aftermath Than the Upheaval" ', *No Depression*, 12 Jul. Available at: http://nodepression.com/interview/willy-vlautin-more-aftermath-upheaval [Accessed 17 Nov.].

Gibson-Graham, J. K. (2006) *A Postcapitalist Politics*. London: University of Minnesota Press.

Goodman, D. (1999) *Modern Twang: An Alternative Country Music Guide and Directory*. Nashville: Dowling Press.

Gusterson, H. (2017) 'From Brexit to Trump: Anthropology and the Rise of Nationalist Populism', *American Ethnologist*, 44(2), pp. 209–214.

Halfacree, K. (2018) 'Hope and Repair Within the Western Skyline? Americana Music's Rural Heterotopia', *Journal of Rural Studies*, 63, pp. 1–14.

Hetherington, K. (1997) *The Badlands of Modernity*. London: Routledge.

Holt, F. (2007) *Genre in Popular Music*. London: University of Chicago Press.
Jones, A. (2009) 'Richmond Fontaine – We Used to Think the Freeway Sounded Like a River', *Uncut September*, p. 80.
Kirby, J. (2006) 'Life a Wrecking Ball: Gillian Welch and the Modern South', MA thesis, Bowling Green State University, OH. Available at: http://rave.ohiolink.edu/etdc/view?acc_num=bgsu1151330052 [Accessed 17 Nov.].
Lamont, M., Park, B. Y. and Ayala-Hurtado, E. (2017) 'Trump's Electoral Speeches and His Appeal to the American White Working Class', *British Journal of Sociology*, 68, pp. S153–S180.
McGrath, K. (2015) 'Interview: Willy Vlautin', *Wales Arts Review*, 27 Nov. Available at: www.walesartsreview.org/interview-willy-vlautin-2/ [Accessed 17 Nov.].
Monnat, S. and Brown, D. (2017) 'More Than a Rural Revolt: Landscapes of Despair and the 2016 Presidential Election', *Journal of Rural Studies*, 55, pp. 227–236.
O'Hagan, S. (2016) 'Willy Vlautin: "I Had a Picture of Steinbeck and a Picture of the Jam"', *Guardian*, 24 Apr. Available at: www.theguardian.com/music/2016/apr/24/willy-vlautin-richmond-fontaine-interview-delines [Accessed 17 Nov.].
Peterson, R. and Beal, B. (2001) 'Alternative Country: Origins, Music, World-View, Fans, and Taste in Genre Formation', *Popular Music and Society*, 25(1/2), pp. 233–248.
Río Raigadas, D. (2016) 'Reinterpreting the American West from an Urban Literary Perspective: Contemporary Reno Writing', in Ibarrola Armendariz, A. and Ortiz de Urbina Arruabarrena, J. (eds.) *On the Move: Glancing Backwards to Build a Future in English Studies*. Bilbao: University of Deusto, pp. 39–50.
Scala, D. and Johnson, K. (2017) 'Political Polarization Along the Rural-Urban Continuum? The Geography of the Presidential Vote, 2000–2016', *Annals of the American Academy of Political and Social Science*, 672, pp. 162–184.
Scoones, I., Edelman, M., Borras, S., Hall, R., Wolford, W. and White, B. (2018) 'Emancipatory Rural Politics: Confronting Authoritarian Populism', *Journal of Peasant Studies*, 45(1), pp. 1–20.
Sherman, J. (2006) 'Coping with Rural Poverty: Economic Survival and Moral Capital in Rural America', *Social Forces*, 85(2), pp. 891–913.
Sherman, J. (2009) *Those Who Work, Those Who Don't: Poverty, Morality, and Family in Rural America*. Minneapolis: University of Minnesota Press.
Sherman, J. (2018) '"Not allowed to Inherit My Kingdom": Amenity Development and Social Inequality in the Rural West', *Rural Sociology*, 83(1), pp. 174–207.
Slatta, R. (2010) 'Making and Unmaking Myths of the American Frontier', *European Journal of American Culture*, 29(2), pp. 81–92.
Statista (2019) 'Music Album Sales in the United States from 2008 to 2014, by Genre (in millions)'. Available at: www.statista.com/statistics/188910/us-music-album-sales-by-genre-2010/ [Accessed 19 Jan.].
Ulrich-Schad, J. and Duncan, C. (2018) 'People and Places Left Behind: Work, Culture and Politics in the Rural United States', *Journal of Peasant Studies*, 45(1), pp. 59–79.
Vlautin, W. (2006) *The Motel Life*. London: Faber and Faber.
Wikipedia (2018) 'Country Music'. Available at: https://en.wikipedia.org/wiki/Country_music [Accessed 18 Dec.].

6 'We Sure Didn't Know'
Laura Gilpin, Mary Ann Nakai, and Cold War politics on the Navajo Nation

Louise Siddons

According to multiple scholars, the aftermath of the Second World War transformed "American Indians" into "Indian Americans" (Bernstein, 1991, p. 159; Morgan, 1995). It is a rhetoric that evokes both the assimilationist politics driving Federal Indian policy throughout the 1950s and the concomitant emergence of tribal nationalisms in that decade – but what does it actually mean? A 1950 photograph of a Navajo family, published by Anglo photographer Laura Gilpin (American, 1891–1979) in her 1968 book, *The Enduring Navaho*, raises precisely this question [Figure 6.1].[1]

In this essay, I ask how Gilpin's photograph queered conversations about authenticity and national identity that were taking place on the Navajo Nation over the course of the mid-twentieth century. Although Mark Rifkin and other Native Studies scholars have already made critiques of authenticity that are explicitly queer, none have come from the perspective of visual studies (Rifkin, 2012). In this essay, I frame the visual politics of "A Navaho Family" within the story of two families: that of Gilpin and her partner, Elizabeth Forster, lesbians who lived in Red Rock, Arizona, Colorado Springs, Colorado, and Santa Fe, New Mexico; and that of Francis Nakai and his wife Mary Ann, who lived in Red Rock, Arizona, and Shiprock, New Mexico – both in the northeast corner of the Navajo Nation.

Although Gilpin has been the subject of significant scholarship, and her portraits of Navajo people have been objects of sustained critical discussion since the 1930s, both histories have consistently been distorted and erased. Gilpin and Forster's lifelong romantic partnership has been described as a "friendship" by the photographer's primary biographer and countless other historians – and no study of the photographs has seriously considered the biographies or the local political context of the Navajo people Gilpin portrayed. The first erasure is a symptom of the heteropatriarchy of the academy; the second is an embarrassing artifact of the distance between "American" and "Native American" studies. The longstanding structures that enable these elisions in the extant scholarship were instrumental in the oppression faced by Gilpin, as a lesbian woman, and by Navajo people, both individually and as a nation, throughout the twentieth century. Addressing these erasures means queering the historical record – not simply adding lesbian or Navajo data points to existing histories, but radically reinterpreting that history

Figure 6.1 Laura Gilpin (1891–1979), *Francis Nakai and Family*, Sept. 17, 1950, gelatin silver print, 10 5/8 × 7 13/16 in., Amon Carter Museum of American Art, Fort Worth, Texas, Bequest of the artist

Source: © 1979 Amon Carter Museum of American Art, P1979.128.509.

from lesbian and Diné perspectives. As art historian Tirza True Latimer suggests, "queer theory . . . enables movement and identification across embodied positions," and "within both queer and feminist studies, scholars face the challenges of studying subjects who have been 'disappeared' from . . . official accounts of the past" (Latimer, 2016, p. 95; Goldberg, 2001). To queer the history of Gilpin's portrait of the Nakai family, then, is simply to tell it for the first time: to see the Nakai family – and Gilpin's construction of their portrait – clearly, and thereby also to see the history of the United States more clearly. It is a strategy that Gilpin herself insisted upon to her readers: "one must know the individuals," she instructed them in the opening pages of her book, prioritizing the experiences of her family and the Nakais over the U.S. government-imposed definitions of Navajo and American authenticity that were being amplified and themselves authenticated by white audiences across the country at mid-century (Gilpin, 1968, p. 23). *The Enduring Navaho* invited readers to shift their perspective on nationalism and Indigenous sovereignty, much as Gilpin shifted her own between 1932 and 1950.

The Navajo Nation is a sovereign Native American nation, with legally defined borders, government, and citizens.[2] Today's political reality is the product of centuries of international conflict and diplomacy – between the Navajo (Diné) and other Indigenous nations, such as the Pueblo and Hopi people, and between the Navajo and the colonial governments of Spain, Mexico, and the United States. Today, the Navajo Nation has jurisdiction over only a portion of Dinétah, the traditional Navajo homeland, and like all Native American tribes within the United States, its sovereign power is limited by its status as a domestic dependent nation within the United States. With over 300,000 citizens (approximately half of whom live outside the Navajo Nation) and more than 27,000 square miles of land, referred to colloquially as Diné Bikéyah or Naabeehó Bikéyah, the Navajo Nation is the largest Indigenous nation in the United States. With the partial exception of their 1864–68 exile to Hwéeldi (also known as Bosque Redondo and Fort Sumner), the Navajo have continuously occupied their traditional homeland.[3] From her home in Colorado Springs, Gilpin traveled through and around Navajo land from the turn of the century onward, although the initial focus of her professional work was on its landscape and regional archaeology, rather than on contemporary inhabitants of the region. In 1931, Gilpin's partner, Forster, accepted a position as a field nurse in Red Rock, a small community in north central Diné Bikéyah, and as Gilpin began to photograph Forster's colleagues and friends, her broader interest in contemporary Navajo life was likewise engaged.

Gilpin built a career out of photographing the American West – its land and its people – for outsider audiences. Before she began making photographs on the Navajo Nation, Gilpin was already implicated in a semiotics of cross-cultural representation informed by histories of colonization and tourism that paradoxically combined genocide and forced assimilation with demands for cultural preservation and picturesque authenticity. Gilpin famously cited a photograph by Edward Curtis (1868–1952) that hung in her childhood home as an early influence, along with the work of her relative, the western photographer William Henry Jackson (1843–1942) (Sandweiss, 1987). Unlike Curtis and Jackson, however, Gilpin spent her life in the Southwest, working around and among the Indigenous, Hispanic and Anglo people of Colorado, New Mexico, Utah, and Arizona. Involved in her community as a business owner, educator, political activist, and advocate as well as an artist, she developed a nuanced understanding of the visual politics of Navajo representation and its connection to the complex sovereignty politics of the Diné throughout the middle of the twentieth century. Over the course of four decades, Gilpin became a sympathetic and politically astute visitor to the Navajo Nation – a perspective that she sought to share with other Anglos in *The Enduring Navaho*. The book is a complex cultural document; like its author, it is both complicit in, and fighting against, the ongoing exploitation that has shaped the Southwestern landscape.

Gilpin was far from alone in turning her attention to the Navajo Nation and its people at mid-century, although the longevity and intensity of her engagement does set her apart from other non-Navajo photographers who were among the many tourists, social scientists, government officials, and others who traveled to Dinétah to document their impressions of its people. In almost every case, they claimed that their products offered an authentic representation of their subjects. Such claims, made by non-Navajo people to primarily non-Navajo audiences, were characteristic of the function of authenticity in Native America. As Jean Fisher and Rae Owens have put it, authenticity is a "function of the [Euro-American] gaze," requiring that "to claim authenticity in the world – in order to be seen at all – the Indian must conform to an identity imposed from the outside" (Fisher, 1992, p. 50; Owens, 2011). Concisely, Scott Richard Lyons has concluded that, "assimilation and authenticity have always been language games designed for Indians to lose" (Lyons, 2011, p. 303). Moreover, discussions of authenticity tend to occlude "tribal diversity, tribal knowledge, and cultural change over time," an erasure that Monica L. Butler observes has "placed a burden on tribes by compelling them to meet standards of cultural authenticity devised by the settler state" (Butler, 2018, pp. 4, 26). Overwhelmingly (and tellingly), academic discussions of authenticity in relation to Native America concern themselves with the experiences, perspectives, and judgments of non-Native people. At the same time, Philip Deloria reminds us that Native people have also participated in constructions of authenticity: "It would be folly," he expostulates, "to imagine that white Americans blissfully used Indianness to tangle with their ideological dilemmas while native people stood idly by, expecting no influence over the resulting Indian images" (Deloria, 1998, p. 8). Oppressive but subject to opportunistic exploitation, the notion of authenticity in Native America is fraught due to its influence on colonial policy and practices, as well as on public opinion via stereotyping (Vizenor, 1999; Ortiz, 1981; Handley and Lewis, 2004; Phillips and Steiner, 1999; Madsen, 2010; Barker, 2011; Garroutte, 2003).[4] In her discussion of Hopi *katsinam tithu*, Pearlstone is particularly attentive to the distance between insider and outsider definitions of authenticity; ultimately, she privileges the former while interrogating the latter (Pearlstone, 2000). Janet Berlo likewise favors in-group definitions of authenticity in her discussion of Navajo sand paintings (Berlo, 2014). In my essay, this tension between insider and outsider interpretations of authentic Diné culture is evident. In her preface to *The Enduring Navaho*, Gilpin disavows her own expertise, writing, "There is no pretense here of a scientific or an ethnologic approach, but all factual statements have been checked with some of our leading scientists and, finally, with the Navaho People themselves." At the close of this introductory text, she repeats that attribution: "I have been fortunate indeed in the friends I have made and the co-operation I have received from interested Navaho People. This, therefore, is an interpretation of a wonderful people just as I have found them" (Gilpin, 1968, p. vii). Underscoring the intersection between her individual interpretation and the authority she acknowledged in her Diné collaborators, Gilpin's book embodies what historian Michael Frisch has termed "shared authority" (Frisch, 1990). In this chapter, I use photographs from

The Enduring Navaho to deconstruct the shift in Gilpin's interpretive stance, from her early tourist-driven work to the final portrait in the book, "A Navaho Family."

One of several group portraits that Gilpin made of the Nakai family over the course of thirty-five years, "A Navaho Family" appears along with several others in *The Enduring Navaho*. Individually, each of the photographs contributes to a specific moment in the book's narrative as Gilpin seeks to describe and defend the sovereign position of the Navajo Nation in relation to contemporaneous American culture. Together, the images offer quiet evidence of the longevity of Gilpin's connection to the Nakai family, and of its contribution to her changing understanding of Navajo politics over the course of the twentieth century. As Larry McSwain, a reporter for the *Gallup Independent*, argued in 1955, when Gilpin took photographs of the same people, decades apart, and juxtaposed those portraits, viewers were presented with "a unique contrast to stereotyped Indian photographs" (McSwain, 1955). Whereas photographs by Curtis or Jackson presented impressions of a "timeless" Indian, Gilpin's sequence of portraits of the Nakais, taken over the course of thirty years, highlights them as individuals looking out at viewers from the present as well as from various points in the recent past. "A Navaho Family" is especially striking in this context, thanks to its inclusion of a United States flag that the family received when their son Luis was killed in the Second World War. The photograph appears in *The Enduring Navaho* as a full-page illustration in an epilogue that explains the circumstances under which it was taken, the longstanding friendship between the photographer's family and the Nakais, and some biographical details about the latter. What did the photograph say about Navajo – and American – identity to postwar and Vietnam-era audiences, and why did that message captivate them? And what does the photograph invite us to discover about Navajo history and politics throughout that period? In partial answer to these questions, I first go back to one of the earliest stories that Forster and Gilpin told about the Nakais.

In 1932, Gilpin and Forster traveled from Red Rock to Santa Fe with their friends Timothy Kellywood and Mary Ann Nakai (née Benally). Kellywood worked as Forster's translator in the small community of Red Rock, Arizona, and Nakai, whom they always knew as "Mrs. Francis," was a neighbor who had likely taught Forster how to weave on a Navajo loom.[5] While on their trip, the group studied Navajo blankets at the newly established Laboratory of Anthropology.[6] "Mr. [Jesse] Nusbaum and Dr. [Harry] Mera spent several generous hours showing us beautiful specimens of blankets from the vaults," wrote Forster to her sister Emily, in a letter that she later revised for publication.

> Mrs. Francis and Timothy were entranced. Timothy repeated again and again in awed wonder, 'We've heard about the old blankets but we sure didn't know that they were as beautiful as this,' and Mrs. Francis touched them all with careful fingers, very evidently making a quiet study of texture and weave, color and design. I am anxious to see the result of such inspiration in her weaving hereafter.
>
> (Forster, 1988, pp. 70–72)

Forster's goal for the visit was aligned with that of the Laboratory staff: Mera, curator of the Laboratory, organized a variety of events for Navajo weavers in order to foster the revival of older styles (Webster, 1996).

Like Mera, Forster privileged the earlier, less elaborate style of Navajo weaving over contemporary fashions, writing to her sister that "Navajo designs have so deteriorated with the demands of our American market that I long to . . . renew old ideals and simplify the elaborations which our Trading Posts display" (Forster, 1988, pp. 70–72). For Forster, Nusbaum, and Mera, the trip to Santa Fe presented an opportunity to connect a contemporary weaver with a more authentic tradition of Navajo weaving – defined by the nurse as predating the "demands of our American market." This desire was not particularly unusual: in fact, their visit to the Laboratory of Anthropology fit in precisely with a broader trend of "[m]useum anthropologists utiliz[ing] their collections to teach young weavers the 'old ways'" (Webster, 1996, p. 423).[7] But while Forster and the Laboratory celebrated the institution's role in cultural preservation, Kellywood spoke directly to its responsibility for Navajo loss.

Kellywood told those assembled "again and again" that he and Nakai (for whom he acted as interpreter), "didn't know that [the old weavings] were as beautiful as this." His repeated protest calls attention to the destruction of cultural memory enacted by Anglo collectors who took material away from Diné communities, reproaching the Laboratory for practices that, counter to their mission of increasing Indigenous access to collections, had significantly impoverished Native communities. Amplifying Kellywood's words, Nakai's sensory interactions with the weavings – "touch[ing] them all with careful fingers" – spoke to the emotion of reunion and reconnection. Although Forster interpreted Kellywood's response as "wonder," and Nakai's physical attention to each weaving as "study," their actions and words can also be read as profound expressions of mourning in the face of cultural loss – or, more specifically, a history of theft that had begun in the nineteenth century.

"[I]t is probable," wrote art critic W. J. Hoffman in 1895, "that more really good examples [of Navajo weaving] can now be seen in New York than in New Mexico." The enthusiasm for Navajo weavings among turn-of-the-century collectors – Hoffman encouraged their use as "admirable covers for divans, portières, or wall-hangings" – was matched by that among anthropologists and museum curators, with predictable results (Hoffman, 1895, p. 119). "A growing worldwide interest in America's native cultures since the late 1800s had resulted in the removal of unknown quantities of paleontological, archaeological, and ethnographic material from Navajoland through the first half of [the twentieth] century," wrote former Navajo Tribal Museum directors Russell P. Hartman and David E. Doyel in their history of the museum (Hartman and Doyel, 1982, p. 241).

> The possibility of establishing a tribal museum to control the loss of these materials, and to preserve them instead in Navajoland for all Navajo people, was discussed as early as 1950, but not until 1961 was the Navajo Tribal Museum actually established.
>
> (Hartman and Doyel, 1982, pp. 241–242)

After a career spent working in and for colonial institutions, Gilpin enthusiastically supported the Navajo Tribal Museum, working with its first director, Martin Link, to donate a set of fifty of her photographs that she specified must be kept and exhibited on the Navajo Nation.[8]

Historian Eric Lott has compellingly documented the ways in which cultural theft, framed in terms of admiration and even love, is at the structural heart of American cultural appropriation (Lott, 1993). The paradox of this structure is clear in Forster's narratives: her affection for Nakai, as well as her love of Navajo weaving, is evident throughout her letters, and yet in the 1930s, both she and Gilpin embraced an institutionalization of Navajo history that was framed by Anglo expertise and authority. At the Laboratory of Anthropology, Nakai was an active witness to this institutional theft of Navajo cultural sovereignty. Although at the time, neither Forster nor Gilpin seems to have perceived their visit to Santa Fe in those terms, by 1968 Gilpin would offer her readers Nakai's experience as witness to U.S. imperialism – and her authority as a Diné woman and family matriarch in the face of its effects on the Navajo Nation – as the powerful culminating image of *The Enduring Navaho*.

Forster was forced to resign her position as a field nurse in Red Rock when the Association on Indian Affairs ran out of funds in 1933. "When the final farewells were said," Forster noted in a letter to her friend Marion, "I saw Mrs. Francis's weeping head descend to Laura's shoulder . . . My heart is truly hurt by the parting" (Forster, 1988, p. 138). Forster and Gilpin left the Navajo Nation, and although they kept up a regular correspondence with Francis Nakai, they did not see the family again until 1950. Many changes had taken place on the Navajo Nation in the intervening years – but one in particular had impacted the Nakai family. Gilpin wrote,

> Still living in the same little house, they then had two daughters, much younger than the boys we had first known. The oldest son [Luis] had been killed in the European theatre of World War II, and almost the only object in the room in which we were sitting was the flag that had been over the boy's coffin at the time of his burial in France.
>
> (Gilpin, 1968, p. 247)

In Gilpin's photograph of the family on this occasion, they all face her camera with unflinching expressions: directing their gazes at the lens, each sitter interrogates Gilpin's gaze and demands that she (and, by implication, all of the photograph's viewers) answers for – and to – the image's disquietude. Disrupting the narrative of a decades-long friendship, "A Navaho Family" conveys instead something of the complex intersections between Navajo identity and American history to a nationwide audience with little or no direct experience of the Navajo Nation, its people, or its values.

The Nationality Act of 1940 paved the way for Native Americans to register for the draft, and ten percent of the Native population served in World War Two – a higher proportion than any other demographic (Ferguson-Bohnee, 2016, p. 1109). Indeed, a 1942 survey found that 40 percent more Native servicemen had enlisted than had been drafted – although they were sometimes led to do so under false pretenses (Morgan, 1995). "[My grandfather] was told if he served, the family would get some of their land back and a house," Navajo airman Philip Rock told interviewer Brye Stevens in 2018. "None of that happened" (Stevens, 2018). Such promises meant that as servicemen returned from the War, their expectations were high – and their disappointment was compounded by a new sense of alienation. Diné historian Alice Bathke and her husband, Jerry, wrote of the war's aftermath,

> When [Navajo servicemen] returned to the reservation they were confused and had serious questions about their identity. Among their families they were looked upon as having lost part of the Navajo way. Sometimes the very worst that can possibly be said of anyone was said to describe their condition: "He acts as if he has no family."
>
> (Bathke and Bathke, 1969, p. 16)

The Nakai family lost their son Luis on a battlefield in Europe; many returning Diné men discovered that even though they had lived through the war, their families had lost them nonetheless.

Thirty-five hundred Navajo men joined the U.S. Army, Navy, and Marines in the Second World War; three hundred lost their lives ("Navajos", 1962). Four hundred of them became "Codetalkers" in the high-profile Marine program that used the Navajo language as the basis for an unbreakable communications code throughout the Pacific theatre, saving thousands of lives. It was by far the most famous contribution of Native people to a U.S. war effort, and one that overshadowed other Navajo service in the popular press (and collective memory), even as it brought Native patriotism into the spotlight. In a 1954 interview, preserved without citation in one of Gilpin's scrapbooks, the photographer discussed Navajo military service in terms of the relationship between the Navajo Nation and the United States and the obligations of each toward the other. "[T]he whole Indian record in the war is superb," Gilpin told the *Denver Post*'s Ellen O'Connor.

> Many of the young men volunteered, some were drafted. They sent home money to aid their families, which was a big help. And they played an important part in winning. Navajos talking Navajo – the Japanese were stuck. Here was a code they couldn't figure out.

But, she noted, the Navajo servicemen's contributions to the war effort and to their families were in stark contrast to the United States' ongoing failure to meet its treaty obligations with the Navajo Nation (O'Connor, 1954).

Two decades earlier, Forster had been prompted by the "bitter wind" of a harsh winter to "unhappy wonder" about the "hardship and want, discomfort and suffering" that surrounded her in Red Rock. "The Navaho," she pointed out, "are wards of our Government. Does that not mean that our Government is responsible for their welfare? . . . Must they die each winter from cold and hunger?" (Forster, 1988, p. 92). In addition to the persistent shortcomings of health and welfare programs across the reservation, which Gilpin and Forster had experienced firsthand, Gilpin reminded O'Connor that "[w]e promised the Navajos by treaty to furnish one teacher for every 30 children, yet today there are 21,000 children who want to go to school and can't." Contrasting Navajo service with U.S. neglect, the photographer chastised the *Post*'s readers for the arrogance of American imperialism. "In all our efforts to bring civilization to the Navajos," she concluded, "[t]here is something which they lose, that we can never give" (O'Connor, 1954). As the Bathkes put it, "The People feel that their development must be completed in the Navajo way . . . they strongly desire the opportunity to determine the course of their own future" (Bathke and Bathke, 1969, p. 17). Throughout the first half of the twentieth century, Navajo leaders repeatedly fought federal policies that promoted assimilation, and pushed back against programs that infringed upon Navajo ways of life.

And yet, "Navajos have always been deeply patriotic," wrote Roseann Sandoval Willink and Paul G. Zobrod in their 1996 catalogue, *Weaving a World*. Like most generalizations, it is an oversimplification that obscures much of the history of Navajo-U.S. relations – but the absoluteness of their statement distills the spirit of many post-World War Two Navajo leaders' rhetoric around citizenship. The year after *The Enduring Navaho* was published, Navajo Tribal Council Chairman Raymond Nakai was presented with a medallion and plaque in honor of the Navajo Codetalkers. Nakai addressed the question head-on: "Many people have asked why we fight the white man's wars. Our answer is always that we are proud to be American, and we are proud to be American Indians." But he did not stop there. As the Vietnam War dragged on, provoking increasing protest across the United States and beyond, Nakai asserted that Navajo people were *more* patriotic than white Americans. "With so many white men burning their draft cards," he proclaimed, "or leading such dissipated lives that they fail their physical exams, the American Indian always stands ready when his country needs him." Finally, the chairman linked that readiness to Indigenous land: "His clean, rigorous life on his reservation keeps him physically fit" ("Tribal Fair", 1969). In a different context, dance historian Theresa Jill Buckland has described "embodied practices" that confer authenticity on their participants; Nakai's emphasis on the physicality of military service, and its connection to the ideal American body, invokes precisely such a structure (Buckland, 2001).[9] Navajo participation in the U.S. military – particularly as the much-lauded Codetalkers – prompted Chairman Nakai and others to draw direct lines from Navajo honor to U.S. nationalism, as well as from Navajo nationalism to U.S. military honor.

But they were still denied American citizenship: as historian C.G. Galloway has documented,

> [m]any expected to return home to a better life and greater equality. Instead, they found . . . continued denial of voting rights that had supposedly been secured by the Indian Citizenship Act of 1924 ("We do not understand what kind of citizenship you would call that," Pvt. Ralph Anderson had written on behalf of "the Navajo soldier boys" in 1943).
>
> (Galloway, 2004, pp. 128–129)

Although the Indian Citizenship Act had declared that "all non-citizen Indians born within the territorial limits of the United States" were from that point forward considered "citizens of the United States," individual states still prevented Native people from voting ("Act", 1924).[10] In New Mexico, for example, the constitution specified, "Indians not taxed may not vote," a loophole that was successfully challenged in the courts but that was not fully closed until the state changed its laws in 1962.[11] The ambivalence of the United States and its Anglo citizens to Navajo Americanness is evident in the celebrity of the Codetalkers: in the same moment, the Navajo language was an American strategic triumph and a language that had been banned in most government-funded schools. In other words, the admiration expressed nationwide for the Navajo Codetalkers was given the lie by that same nation's ongoing policies of genocide and assimilation against the Diné. The contrast between Navajo men's willingness to serve in the U.S. military and their ongoing disenfranchisement is similarly thematized in "A Navaho Family," as Gilpin uses the U.S. flag to separate the only adult male in the photograph, Francis Nakai, from his wife and the children seated below it. Gilpin's composition aligns Navajo masculinity with U.S. nationalism, which oppresses the rest of the family – the central unit of Navajo political, as well as personal, life – beneath the flag's stripes.

As Diné historian Jennifer Nez Denetdale has warned, rhetoric that equates Navajo military service with Diné patriotism threatens to "streamline Native pasts into the dominant American narrative about itself as a multicultural nation founded on moral and ethical principles and erase the historical links between the past and the present." In other words, by equating Navajo national pride with that of United States citizens more broadly, Raymond Nakai was in danger of retroactively excusing (and even offering moral justification for) the horrors of, for example, Kit Carson's U.S. military campaign against the Navajo, which culminated in their exile to Hwéeldi. Such atrocities, moreover, are not restricted to the past: "violently dispossessed of most of their lands," Denetdale points out, Native people have seen "their sovereign statuses as nations continually undermined by U.S. federal Indian policies and the Supreme Court" (Denetdale, 2009, p. 131). Challenges to sovereignty at the national level parallel Navajo disenfranchisement at the individual level, as they are part and parcel of the same colonial structure. In "A Navaho Family," Gilpin juxtaposes the anonymous, institutional symbolism of the U.S. flag with the individual members of the Nakai family – but also complicates that opposition, because in this situation, the flag itself stands for both colonial nation and Navajo son. Arrayed as the others are along the blank wall of the "small frame house" in which the Nakais had lived for over two decades, the flag in Gilpin's photograph is one more member of the family.

That flag is not wholly visible: Gilpin included just its lower seven stripes and a fragment of the bottom row of stars in her frame. Because her print is black and white, this slender, cropped row of stars is the most immediate clue that the hanging textile is, indeed, the American flag. Thus cropped, the lower register of the flag bears striking formal similarity to the horizontal stripes of nineteenth-century Navajo blankets – the very type of blankets by which Nakai and Kellywood were confronted in the Laboratory of Anthropology, and whose loss they felt so deeply. The Laboratory (now part of the Museum of Indian Arts and Culture) holds "some of the earliest Navajo textiles in existence, dating from 1750 to 1803, and includes a large collection of exemplary Navajo blankets from the 19th century" (Museum, 2019). Between approximately 1820 and the Hwéeldi period, Navajo weaving was characterized by formal simplicity. "First-phase chief's blankets," include only horizontal stripes woven in the "traditional palettes of red, blue, black, and white" (Webster, 1996, p. 428). Called "chief's blankets" because they were affordable only to the most prestigious and powerful leaders of the Navajo and other tribes, such weavings were symbols of status and power. Several examples of this style of blanket are at the Laboratory of Anthropology: for example, a Ute-style first-phase blanket from 1850, originally owned by Ouray, chief of the Uncompaghre Ute and his wife, Chepita, has six alternating dark red-brown and white stripes on either side of a central block of alternating blue and brown stripes [Figure 6.2]; an 1860 blanket incorporates a brighter red into a similar palette of dark blue, dark brown, and white [Figure 6.3].[12]

In "A Navaho Family," Gilpin's framing emphasizes the semiotic slippage between the flag, the body of Luis Nakai – and a woven blanket stylistically congruent with the "old blankets" that Kellywood and Nakai "didn't know." Did Mary Ann Nakai recognize the U.S. flag she received after her son's death in the same terms as her encounter with blankets at the Laboratory of Anthropology? Seen thus, the flag becomes the material evidence of the theft of her son

Figure 6.2 Unidentified Navajo artist, *Phase I Chief Blanket*, 1800–1850, Museum of Indian Arts & Culture/Laboratory of Anthropology, 9117/05.

Figure 6.3 Unidentified Navajo artist, *Phase I variation Chief Blanket*, 1850–1865, Museum of Indian Arts & Culture/Laboratory of Anthropology, 9118/01.

by a nation that refused fully to acknowledge Luis as one of its own, even as it demanded his service.

Once seen as a blanket, the flag is no longer just a symbol of cultural authority and military service. For Mary Ann Nakai and other Navajo women, weaving represented family relationships, economic power, and Navajo sovereignty. Sheep, and the products produced using their wool, were historically at the heart of Navajo economics – but the pastoral economy was severely disrupted by the 1930s stock reduction program. In the 1954 interview cited above, Gilpin blamed the "deplorable" living conditions in Navajo communities on John Collier's disastrous stock reduction program of the 1930s: "Their whole economic system was upset with the sheep reduction program," she asserted, implicating that policy, and U.S. government mismanagement generally, in the ongoing paucity of healthcare and education on the Navajo Nation – as well as in the destruction of "a way of life that had its own perfections" (O'Connor, 1954). In the 1920s and 1930s, the debate over sheep – and, by direct extension, over weaving – was a key component of Navajo sovereignty, central to arguments about U.S. interference with the Navajo because of its intersection with gender politics, domestic assimilation and education, the emergence of wage labor, stock reduction and breeding programs, and more (Weisiger, 2009). In this context, the flag in "A Navaho Family" is a reminder not only of the Nakais' loss of Luis but also of the widespread loss of matriarchal authority on the Navajo Nation as a result of U.S. imperialism.

'We Sure Didn't Know' 87

Art historians have only recently begun attending to the economic, cultural, and political significance of Navajo weaving, rather than investigating its formal qualities in the service of connoisseurship. Formalist approaches to the study of Navajo weaving were in keeping with the preoccupation with authenticity that motivated dealers – and Forster – in their attempts to revive nineteenth-century textile designs among twentieth-century weavers. In her revolutionary feminist analysis of the economic impact of weaving on the Navajo Nation, historian Kathy M'Closkey noted that Navajo women "retained complete control of" their textiles rather than ceding economic power to Navajo men as white traders had initially anticipated. Women traded their own weavings and held their own accounts, using their income to provide vital contributions to their households (M'Closkey, 1994, p. 215). "Decade after decade," M'Closkey emphasized, "the survival of the Navajo rested primarily on women's textile production and the wool from their sheep" (M'Closkey, 1994, p. 216). The historian suggests that the economic significance of Navajo weaving has been downplayed because it was controlled by women; a form of discrimination familiar to both Forster and Gilpin as they built their own careers. It is not surprising, then, that when she wrote to her sister describing the 1932 visit to Santa Fe, Forster not only acknowledged but foregrounded the role played by weaving's economic power in its aesthetic development.

In that letter, Forster's use of the second-person plural-possessive ("our American markets") implicitly included Navajo weavers alongside the proprietors of trading posts and their tourist customers. In other words, Forster's grammar interpellates the weavers as "American" via their participation in the nation's capitalist market – even though for Kellywood and Nakai, the visit to the Laboratory of Anthropology may well have underscored their oppositional relationship to Americanness and to the colonial history that destroyed, appropriated, and then belatedly claimed to preserve Navajo culture. That so-called preservation occurred in the context of a salvage archaeology whose practitioners privileged an ahistorical past that they characterized as authentic, traditional, and timeless – even when they were collecting contemporary blankets, and despite their dependence upon weavers' ongoing presence and production. Whether she intended it or not, Forster's grammar acknowledges the centrality of the weavers themselves to the development of the market for their products – something that few commentators in the 1930s recognized.

The economic value of weaving is visible in a series of portraits that Gilpin made of Nakai and her sons, Juan and Luis, in 1932, just before they set off on their trip to Santa Fe. In all of the photographs, the young mother is wrapped in a blanket – but not one that she wove herself [Figure 6.4].[13]

Instead, it is a Pendleton blanket, in all likelihood purchased with proceeds from Nakai's own weaving (M'Closkey, 1994, pp. 188, 204).[14] Throughout the first half of the twentieth century, Pendleton blankets were indelibly associated with Native American culture and aesthetics; as Hartman and Doyel note, "the Navajo have become so closely identified with [Pendleton blankets] that they might be considered part of the native material culture, were it not for their actual origins" (Hartman and Doyel, 1982, p. 252). Thus, although Gilpin was in some

88 *Louise Siddons*

Figure 6.4 Laura Gilpin (1891–1979), *Mrs. Francis Nakai and Son*, 1932, gelatin silver print, 8 1/2 × 7 1/16 in., Amon Carter Museum of American Art, Fort Worth, Texas, Bequest of the artist

Source: © 1979 Amon Carter Museum of American Art, P1979.128.82.

sense simply recording everyday fact when she photographed Nakai wearing one, she was also aware of the iconic quality – and market value – of a photograph of a beautiful Navajo woman wrapped in a Pendleton blanket (Babcock, 1990). This is evident in the series of photographs of the Nakai family that she took that day in 1932: although some are intimately candid and informal, most of the portraits of Mary Ann Nakai are framed closely around her head and shoulders, her expression stereotypically serious. In some, Gilpin used raking light to call attention equally to the young woman's silver jewelry and the contours of her face. The photographer sold one of these more formal portraits as a postcard, titled simply, "Navaho Indian Woman," throughout the rest of her career [Figure 6.5].

"Navaho Indian Woman" depicts Nakai in profile, her hair pulled back in traditional style and held in place with two bobby pins. Her exposed ear has an earring, and various other items of jewelry are visible at Nakai's neck, between the closely held folds of the blanket. Her face is expressionless, and the portrait has all the atmosphere of a mug shot or a phrenological study – contexts that are, indeed, problematic frames for the image in its public role as representative depiction of a Navajo woman, rather than a portrait of an individual. Offering Nakai as an ethnographic type alongside several others in the set of postcards of Navajo subjects, Gilpin's use of profile underscores the extent to which Nakai is reduced to an object of a touristic (implicitly white) gaze, without any possibility

Figure 6.5 Laura Gilpin (1891–1979), *[Mrs. Francis Nakai]*, 1932, gelatin silver print, 9 11/16 × 7 3/4 in., Amon Carter Museum of American Art, Fort Worth, Texas, Bequest of the artist

Source: © 1979 Amon Carter Museum of American Art, P1979.128.711.

of returning that gaze. Did Gilpin understand the dangerous politics of "Navaho Indian Woman"? Surely, the answer in 1932 was negative – but over time, her understanding deepened.

Anthropologist M. Jill Ahlberg-Yohe has noted that despite extensive recent scholarship on the market for Navajo weaving, "[f]ar less attention has been paid to local conceptualizations of the exchange systems within which these transactions take place or to Navajo weavers' perspectives on the exchange of their weavings" (Ahlberg-Yohe, 2008, p. 368). Forster did not record whether or not she asked Nakai directly about her experience at the Laboratory of Anthropology, and despite my suggestion that we read the weaver's reaction differently than Forster did, it is impossible to deduce with any certainty Nakai's opinions about her weaving from extant records. Yet Ahlberg-Yohe's interviews with weavers working seventy years after Nakai's visit to the Laboratory of Anthropology suggest a possible scope within which the latter understood her interactions with the weavings in the collection. "My mother told me these things," one of Ahlberg-Yohe's informants told her. "Never forget the sheep. *Dibé, dibé*, they will feed you. They clothe you. They bring you things." Later, the same woman underscored her comment: "Remember what I told you about that one time? It is *naalyéhé*. Weaving brings things, the naalyéhé, it brings things like the sheep, the jewelry, the food. You get things from weaving" (Ahlberg-Yohe, 2008, p. 370).[15] This exchange is not simply economic or transactional, however: rather, it

> involves notions and acts of reciprocity, respect, and exchange beyond the marketplace . . . Weaving is tied to a way of viewing the world . . . Selling a weaving is not only a viable source of income, it is also part of a cosmological exchange among Navajo weavers and the Holy People, or the *Diyin Dine'é*, and the world around them.
>
> (Ahlberg-Yohe, 2008, p. 370)

By 1950, Gilpin was beginning to understand the significance of weaving to her friend, and to the Navajo people more broadly – and by the time she planned the final version of *The Enduring Navaho* a decade or so later, she chose to include a different portrait of Mary Ann Nakai.

In an explicit rejection of the politics of "Navaho Indian Woman," the portrait Gilpin included in *The Enduring Navaho* is her candid depiction of Nakai in the company of her younger son, Juan, rather than the postcard portrait. If "Navaho Indian Woman" encapsulated the settler-colonial gaze, then the double portrait of Mary Ann and Juan Nakai reveals the intimate gazes of friends and family. Her friendships with individual Navajo people spurred Gilpin's shift from a paternalistic, romanticizing vision of Native American culture to a more contemporary understanding of Navajo politics. The photographer's decision to use her double portrait of Mary Ann and Juan alongside "A Navaho Family" in the book's epilogue was in keeping with her intent to make the case for Navajo sovereignty. Unlike the stereotypical stoicism of the postcard photograph, both mother and child in this image appear distracted by something – or someone – entertaining them beyond the frame. Nakai's mouth is turning up in the beginning of a smile, and Juan's more skeptical curiosity suggests that perhaps his elder brother is the one whose off-camera antics are more compelling than the black box and lens of Gilpin's large-format camera. The wall behind Nakai and her son is utterly blank, giving it the appearance of a studio portrait, even though Gilpin's description makes it clear that they are at home, in the "small frame house directly south of Betsy's quarters, perhaps a quarter of a mile away" (Gilpin, 1968, p. 247). This proximity was doubtless a factor in the women's friendship, as Gilpin noted that Nakai

> was one of Betsy's frequent visitors. She would come, sit for a while, drink a cup of coffee, but, as she spoke no English, conversation was very limited. She had great charm, however, and whenever I paid Betsy a visit I saw her frequently.
>
> (Gilpin, 1968, p. 247)

Unlike the profile portrait that Gilpin had used for her postcard, the photograph of Nakai and her son reproduced in *The Enduring Navaho* reveals this charm; we can easily imagine all present dissolving into laughter seconds after the shutter was released.

In the book, the portrait of Mary Ann and Juan is on the upper left-hand side of a page spread, facing "A Navaho Family," with a color plate tipped in between. Below the mother and son is a portrait of Francis Nakai, alone, from the same year, and alongside these two photographs, on the same page, Gilpin's text describes Francis Nakai's death and Forster's nostalgia as she contemplates the loss of her friend alongside changes in the Red Rock landscape. It is a poignant moment in the book's narrative, made more somber by the drawn faces of the Nakais in "A Navaho Family" on the facing page. Seen side by side, the stripes of Mary Ann Nakai's Pendleton blanket in 1932 very obviously echo those of the American flag in 1950, adding a political edge to the passage of time revealed by the two versions of her face. Together, these juxtapositions invite us to scrutinize the two images – and particularly, Mary Ann Nakai, who from one photo to the other turns her attention from her children to the camera and the woman standing behind it. In so doing, Gilpin intentionally unsettles expectations of static, timeless Indianness – the colonizing codes of "authenticity" – and instead foregrounds her own changing subjectivity alongside the physical changes undergone by her subject.

Each of the five family members included in "A Navaho Family" wears a slightly different expression – Gilpin characterizes Francis and Mary Ann as "dejected," whereas the three children's gazes have an uncompromising quality about them – but all look directly at us. Their bodies spiral around the photograph, trailing from the lower left up, across the flag, down Francis Nakai's body, back up through the younger girl, to the seated matriarch at the center of the photograph. Mary Ann Nakai herself sits in a rocking chair, flanked by the two young girls whose hands cradle her chair and knee protectively as they confront Gilpin's camera. Uncanny doubles of Nakai's absent sons, they remind us (along with Gilpin's text) of all that Mary Ann Nakai has lost – but they also represent her future. Reaching out from the covered ground of a flag that figures their son's body and rhymes with pre-exile-era Navajo weaving, the Nakai family's unsettling gazes challenged viewers expectations about authenticity and Navajo experience, disrupting the complacency of settler colonialism, patriarchy, and American nationalism during the Cold War (Horton, 2017, p. 131; Durham and Fisher, 1988).

Notes

1 "Anglo" was Gilpin's descriptor of choice for white, English-speaking Americans – including herself.
2 The Navajo Nation was only formally designated as such in 1969. In the 1930s, many people used "Navajoland" to describe the Navajo Indian Reservation; today, several designations are used alongside Navajo Nation.
3 Some Navajo resisted Kit Carson's brutal campaign and the exile to Hwéeldi, aided by Apache people with whom they later joined to become the Alamo chapter of the Navajo Nation, with their own (non-contiguous) reservation land.
4 An additional body of scholarship on Native authenticity addresses questions of legislative authentication of Native identity, using blood quantum, etc. Although it is beyond the scope of this essay, it is worth noting that the name Nakai means "Mexican" in Diné bizaad (the Navajo language), hinting at the complexity of the ethnohistory of the American Southwest.
5 Naming among the Diné is complicated; in the 1930s, many Diné considered it inappropriate to refer to someone directly by name, and so nicknames and relational references were common. Mrs. Nakai spoke no English, and appears to have found "Mrs. Francis" to be an apt and functional relational reference name to be used by Betsy and Laura. Over time, conventions changed; by 2009, when the Nakais' great-granddaughter, Marguerite Sheehan, was celebrated as her high school salutatorian in the *Navajo Times*, Mrs. Nakai was listed as Mary Ann Benally ("Education Briefs," *Navajo Times*, August 13, 2009).
6 Gilpin and Forster's association with the Laboratory of Anthropology likely reached back before its founding, since Gilpin had known Jesse Nusbaum since his tenure as superintendent of Mesa Verde National Park, where she was commissioned to take photographs in 1925 (their long friendship with the Nusbaums may have been one of several that drew Gilpin and Forster to relocate to Santa Fe after the Second World War).
7 Somewhat ironically in light of Forster's identification of the market as a corrupting force, art historian Karen Ohnesorge has pointed out that authenticity itself is a marketing concept anchored in the notion of the traditional: "authenticity," she writes, "is still marketed to the public as a matter of conformity to traditionalism" (Ohnesorge, 2008, p. 53).
8 Drafts of this agreement are in the Gilpin papers at the Amon Carter Museum of American Art. Link was also a founder, during his time at the Museum, of the Navajo Code Talker Association.
9 Buckland notes that controlled access to disciplined spaces of embodiment are key to this perception of authenticity: discussing the English folk dance known as Abbots Bromley, she observes "the distinction of the authentic person through knowledge and expertise acquired not through the inscribing process of books, available to all, but through the physicality of dance, access to which was controlled" (Buckland, 2001, p. 13).
10 The assimilationist 1887 Dawes Act had extended citizenship to any Native person who had "voluntarily taken up . . . his residence separate and apart from any tribe of Indians . . . and has adopted the habits of civilized life," but this did not apply to most Navajo people.
11 Note that voter suppression remains a problem throughout Indian Country, including on the Navajo Nation (Brewer and Petersen, 2018).

12 Both of these blankets were donated to the Laboratory by Frances Stewart, wife of well-known Colorado statesman Philip Stewart and well-known collector of historic Native American art. The Stewarts lived in Colorado Springs from 1900 onward – where it seems likely that, given their shared interests, they were in the same social network as Gilpin and Forster (Stiver, 2000).
13 Interestingly, another postcard in the series, titled "Navaho Medicine Man," is a very similar portrait: a close-up of Tsetah Begay, also of Red Rock, wrapped in a Navajo-woven blanket that covers all but his eyes and nose. We can only speculate about why his blanket was traditionally, rather than commercially, produced. Perhaps Begay felt that a traditional blanket was more appropriate for his work – or it may not even have been his, as in the rest of the portraits Gilpin took of Begay in 1932, he has no such garment, instead wearing a traditional masculine headwrap, button-down shirt, and trousers.
14 Portland, Oregon-based Pendleton Woolen Mills was founded in 1909 by a longtime mill worker, Thomas Lister Kay, who had previously operated the Thomas Kay Woolen Mill, opened in 1889. Originally trading blankets with the Nez Perce people of the Columbia River valley, they expanded into the Southwest, creating specialized designs to appeal to Navajo, Hopi, and Zuni taste. In the words of Cherokee scholar Adrienne Keene, "There's just something distinctly *Native* about Pendleton," even though it is not a Native company (Keene, 2011).
15 *Dibé* is Diné Bizaad for "sheep," and *naalyéhé* is Diné Bizaad for material goods.

Reference list

Act of Jun 2, 1924, Pub. L. No. 68–175, 43 Stat. 253 (1924).

Ahlberg-Yohe, M. (2008) 'What Weavings Bring: The Social Value of Weaving-Related Objects in Contemporary Navajo Life', *Kiva*, 73(4), pp. 367–386.

Babcock, B. (1990) '"A New Mexican Rebecca": Imaging Pueblo Women', *Journal of the Southwest*, 32(4), pp. 400–437.

Barker, J. (2011) *Native Acts: Law, Recognition, and Cultural Authenticity*. Durham, NC: Duke University Press.

Bathke, J. and Bathke, A. (1969) 'They Call Themselves "The People"', *The University of Chicago Magazine*, 61(5), pp. 2–17.

Berlo, J. (2014) 'Navajo Sandpainting in the Age of Cross-Cultural Replication', *Art History*, 37(4), pp. 688–707.

Bernstein, A. (1991) *American Indians and World War II: Toward a New Era in Indian Affairs*. Norman, OK: University of Oklahoma Press.

Brewer, G. and Petersen, H. (2018) 'In Southern Utah, Navajo Voters Rise to Be Heard', *High Country News*, 28 Oct. Available at: www.hcn.org [Accessed 3 Aug. 2019].

Buckland, T. (2001) 'Dance, Authenticity, and Cultural Memory: The Politics of Embodiment', *Yearbook for Traditional Music*, 33, pp. 1–16.

Butler, M. (2018) '"Guardians of the Indian Image": Controlling Representations of Indigenous Cultures in Television', *American Indian Quarterly*, 42(1), p. 1–42.

Deloria, P. (1998) *Playing Indian*. New Haven, CT: Yale University Press.

Denetdale, J. (2009) 'Securing Navajo National Boundaries: War, Patriotism, Tradition, and the Diné Marriage Act of 2005', *Wicazo Sa Review*, 24(2), pp. 131–148.

Durham, J. and Fisher, J. (1988) 'The Ground Has Been Covered', *Artforum International*, 26(10), pp. 99–105.

'Education Briefs' (2009) *Navajo Times*, 13 Aug., p. 24.

Ferguson-Bohnee, P. (2016) 'The History of Indian Voting Rights in Arizona: Overcoming Decades of Voter Suppression', *Arizona State Law Journal*, 47, pp. 1099–1144.

Fisher, J. (1992) 'In Search of the "Inauthentic": Disturbing Signs in Contemporary Native Art', *Art Journal*, 51(3), pp. 44–50.

Forster, E. (1988) *Denizens of the Desert: A Tale in Word and Picture of Life Among the Navaho Indians, the Letters of Elizabeth W. Forster*. Edited by Martha Sandweiss. Albuquerque, NM: University of New Mexico Press.

Frisch, M. (1990) *A Shared Authority: Essays on the Craft and Meaning of Oral and Public History*. Albany, NY: State University of New York Press.

Galloway, C. (2004) 'Endurance and Photography in Navajo History', *Pacific Historical Review*, 73(1), pp. 127–132.

Garroutte, E. (2003) *Real Indians: Identity and the Survival of Native America*. Berkeley, CA: University of California Press.

Gilpin, L. (1968) *The Enduring Navaho*. Austin, TX: University of Texas Press.

Goldberg, J. (2001) *Willa Cather and Others*. Durham, NC: Duke University Press.

Handley, W. and Lewis, N. (eds.) (2004) *True West: Authenticity and the American West*. Lincoln, NE: University of Nebraska Press.

Hartman, R. and Doyel, D. (1982) 'Preserving a Native People's Heritage: A History of the Navajo Tribal Museum', *Kiva*, 47(4), pp. 239–255.

Hoffman, W. (1895) 'Native American Blanket-Making', *The Monthly Illustrator*, 4(12), pp. 114–119.

Horton, J. (2017) *Art for an Undivided Earth*. Durham, NC: Duke University Press.

Keene, A. (2011) 'Let's Talk About Pendleton', *Native Appropriations*. web log post, 3 Feb. Available at: https://nativeappropriations.com/2011/02/lets-talk-about-pendleton.html [Accessed 3 Aug. 2019].

Latimer, T. (2016) 'Improper Objects: Performing Queer/Feminist Art/History', in Jones, A. and Silver, E. (eds.) *Otherwise: Imagining Queer Feminist Art Histories*. Manchester: Manchester University Press.

'Little Promoted Code Talker Museum' (2012) *Navajo Times*, 5 Jan. Available at: www.navajotimes.com/ [Accessed 3 Aug. 2019].

Lott, E. (1993) *Love and Theft: Blackface Minstrelsy and the American Working Class*, Oxford: Oxford University Press.

Lyons, S. (2011) 'Actually Existing Indian Nations: Modernity, Diversity, and the Future of Native American Studies', *American Indian Quarterly*, 35(3), pp. 294–312.

Madsen, D. (ed.) (2010) *Native Authenticity: Transnational Perspectives on Native American Literary Studies*. Albany, NY: State University of New York Press.

M'Closkey, K. (1994) 'Marketing Multiple Myths: The Hidden History of Navajo Weaving', *Journal of the Southwest*, 36(3), pp. 185–220.

McSwain, L. (1955) 'Noted Photographer Compiles Navajo Book', *Gallup Independent*, 26 Jan., p. 1.

Morgan, T. (1995) 'Native Americans in World War II', *Army History*, 35, pp. 22–27.

Museum of Indian Arts and Culture (2019) Santa Fe, NM. Available at: http://indianartsandculture.org/ [Accessed 3 Aug. 2019].

'Navajos – Past, Present and Future' (1962) *Navajo Times*, 5 Sept, p. 11.

O'Connor, E. (1954) 'Laura Gilpin Top Photographer and Writer on S. W. Indian Life', *Denver Post*, 15 Mar., unpaginated.

Ohnesorge, K. (2008) 'Uneasy Terrain: Image, Text, Landscape, and Contemporary Indigenous Artists in the United States', *American Indian Quarterly*, 32(1), pp. 43–69.

Ortiz, S. (1981) 'Toward a National Indian Literature: Cultural Authenticity in Nationalism', *MELUS*, 8(2), pp. 7–12.

Owens, R. (2011) 'Mapping a History: The Zuni Map Project Explores a Culture of Stories and Places', *Flagstaff Live*, 14–20 Apr. Available at: www.flaglive.com/flagstafflive_story.cfm?storyID=222136 [Accessed 16 Aug. 2011].

Pearlstone, Z. (2000) 'Mail-Order "Katsinam" and the Issue of Authenticity', *Journal of the Southwest*, 42(4), pp. 801–832.

Phillips, R. and Steiner, C. (1999) *Unpacking Culture: Art and Commodity in Colonial and Postcolonial Worlds*. Berkeley, CA: University of California Press.

Rifkin, M. (2012) *The Erotics of Sovereignty: Queer Native Writing in the Era of Self-Determination*. Minneapolis, MN: University of Minnesota Press.

Sandoval Willink, R. and Zolbrod, P. (1996) *Weaving a World: Textiles and the Navajo Way of Seeing*. Santa Fe, NM: University of New Mexico Press.

Sandweiss, M. (1987) 'Laura Gilpin and the Tradition of American Landscape Photography', in Norwood, V. and Monk, J. (eds.) *The Desert Is No Lady: Southwestern Landscapes in Women's Writing and Art*. New Haven, CT: Yale University Press.

Stevens, B. (2018) *Navajo Airman Continues Family's Military Legacy at Whiteman*, DF.mil. Available at: www.usafe.af.mil/News/Article-Display/Article/1710269/navajo-airman-continues-familys-military-legacy-at-whiteman/ [Accessed 3 Jan. 2019].

Stiver, L. (2000) 'The Stewart Family', *Keystone of the Arch: The Stewart Collection*. Available at: www.miaclab.org/exhibits/stewart/family.html [Accessed 5 Aug. 2019].

'Tribal Fair One of Best' (1969) *Navajo Times*, 4 Sept., p. 1.

Vizenor, G. (1999) *Manifest Manners: Narratives on Postindian Survivance*. Lincoln, NE: University of Nebraska Press.

Webster, L. (1996) 'Reproducing the Past: Revival and Revision in Navajo Weaving', *Journal of the Southwest*, 38(4), pp. 415–431.

Weisiger, M. (2009) *Dreaming of Sheep in Navajo Country*. Seattle: University of Washington Press.

7 Opening the memory boxes

Magical hyperreality, authenticity and the Haida people

Jane Lovell

Introduction

This chapter explores how perceptions of the "magically real" authenticity of Haida art and culture and the liminal landscape of the Haida Gwaii archipelago have emerged over time. Myths concern the origins of a people, often intertwined with a supernatural element and their key characteristic is that they are continuously adapted (Barthes, 1957). "Place myths" can be generated in tourism (Shields, 1990) by interweaving the history of places with the symbolic to create storied landscapes. The article makes an original contribution to authenticity research by examining how the *Super, Natural* branding of British Columbia offers a multi-layered simulacrum of a magically real place. The brand frames Haida Gwaii for visiting tourists, who encounter "magical" vernacular attributes such as the peripheral location and atmospheric natural landscape, the immanent ghost towns and ruins, material culture and art imbued with the relationship of the Haida people to wildlife and land, and the powerful intangible history of loss and resilience.

The Indigenous Tourism Association of Canada's Economic Impact of Aboriginal Tourism in Canada 2015 estimated that the Aboriginal tourism industry in Canada employs more than 33,000 people and produces $1.4 billion toward Canada's annual GDP (National Aboriginal Tourism Research Project, 2015). Of this national picture, the Haida Gwaii Community Tourism Strategy (2009, p. 16) delineates that Haida Gwaii receives 50,000 visitors a year, with proportionally more international visitors than the rest of British Columbia. However, the sociocultural impacts of tourism include the "place myths" which are established to promote Haida Gwaii. As Cooke (2027, p. 36) argues of the resort of Sun Peaks, "By way of these physical, structural, and discursive displacements and dispossessions of Indigenous peoples and epistemologies, settler colonial places emerge as singular and normative" and also, "morally loaded". The chapter explores the presentation of an essentialist, single, magically real, Haida Gwaii in tourist literature at international, federal and provincial levels using a discourse of authenticity. In parallel with discourses of intellectual decolonialisation, for example the renaming of the Queen Charlotte Islands as Haida Gwaii in 2010, images of Haida art circulate and are copied online, creating a debate about dispossession, appropriation, the meaning of authenticity and decolonialism in the age of digitisation and hyperreality. The chapter indicates the problematic concept that when beliefs about place are elevated to a regional and arguably international level, they become a simulacrum.

Tourism provides the ideal crucible in which to examine authenticity because of both the place marketing theories and the direct, unscripted encounters between hosts and travellers. In particular, "staged authenticity" (MacCannell, 1973, p. 105) is consumed as authentic by visitors, but is argued to be an imitation, curated for consumption. MacCannell (1976, p. 105) posited that host communities provide a seemingly "staged backstage", drawing the appropriating gaze away from the actual, private backstage. Staged authenticity can distort the perspective of places (Lovell and Bull, 2017, p. 17) such as Haida Gwaii, making it "auratic"; not simply physically remote, but unreachable. (Benjamin, 1936). Building on the concept of staged authenticity, this research explores the different stages that are created at governmental and local levels. The work is interdisciplinary, linking human geography, tourism, heritage, indigenous and Canadian studies, and has an anthropological and cultural emphasis.

The history of Haida Gwaii's cultural placemaking

Haida Gwaii is a peripheral edgeland on the Northwest Pacific Coast, which receives 50,000 visitors a year, according to the Haida Gwaii Community Tourist Strategy (nd). Myths are said to be the stories of the origin of a people, and a formative and pivotal event that created a psychological fault line in the origin of the contemporary Haida Gwaii was the accelerated impact – cultural boom followed by collapse – of encounters with Europeans. Following successful sea otter trading, diseases such as measles and smallpox (particularly in 1862) caused the population to drop dramatically (Dowie, 2017, p. 20). Most island villages were abandoned by 1890 and the population moved to the villages of Skidegate and old Massett. The abandoned villages, including longhouses and mortuary poles, have been partially reclaimed by the landscape, a highly distinctive, evocative mix of ancient cedar forest and moss, and are strictly conserved and protected as cultural heritage environments. Despite the dramatic fall in their population, the Haida are known for cultural resilience demonstrated by perpetuated crafts and traditions. The unceded Queen Charlotte Islands were renamed Haida Gwaii in 2010 as a statement of Indigenous ownership and local people have won a long-fought (and rare) achievement of "aboriginal title" which enables then to decide through tenure who can live on the land. The cultural resilience of the Haida militates against narratives of place and dispossession which can occur as indigenous people are "disappeared from textbooks and media representations" (Todd, 2016, p. 4) during the process of colonialisation. This paper indicates that far from having disappeared, media representations internationally promote indigenous culture as the embodiment of supernatural properties, paradoxically staging authentic lives as magically real, the people vanishing behind the myth, making remoteness more than a matter of physical location.

98 *Jane Lovell*

The Pacific Northwest has been popularised by Europeans since the 1870s, when it could be argued that constructions of Haida authenticity truly began to shape public perceptions of local people and their art. Aesthetic appreciation and scientific impetus were compounded by a fear of the loss of indigenous people and their art, which had been realised in the devastated population of Haida Gwaii. George Dawson's 1880 geological expedition encouraged the myth of the "vanishing Indian" (Grek-Martin, 2007, p. 373) and in 1898, Julian Ralph described Indians as a "dead but unburied race" (Ralph cited in Jenkins, 2004, p. 73). Seeing the smouldering ruins of a medicine lodge, Walter McClintock concluded that "in a few years, the ancient customs and ceremonies of the Indians will have disappeared as completely" (Jenkins, 2004, p. 74). Jenkins made the point that that North Americans became engaged in discovering their aboriginal culture after the closure of the Frontier. Writings about the "Red Indian imagination" were visible from the 1890s and folklorists collected examples of the dance and music, which tapped into a diffuse sense of nostalgia for the rural idyll in an era of rapid industrialisation (Lovell and Bull, 2017, p. 17), all of which accelerated the desire to encounter the pre-modern.

Signifiers of Haida material culture were established during colonial encounters with anthropologists such as Franz Boas: "Boas was clearly concerned with the age and authenticity of the objects which he collected" (Jonaitis, 1992, p. 38). The artefacts he amassed included work such as Charlie Edenshaw's, which in effect defined what Haida art was at that time (1897) (Berlo, 1992, p. 5). The intention of Boas when he wrote *Indian Tribes of the North Pacific Coast* in 1895 – a work meticulously illustrated with Chilkat blanket designs, cedar bark hats shapes and other vernacular items – was to give the artefacts meaning, celebrating "primitive" Northwest coastal skills. Cohen states that

> authenticity, for many curators and ethnographers, is principally a quality of pre-modern life, and of cultural products produced prior to the penetration of modern Western influences: hence the common emphasis on cultural products which were "handmade" from "natural" materials.
> (Cohen, 1988, p. 375)

The art symbolised the authenticity of self: "individual freedom and self-realisation, equality and paradise" (Hennig, 2002, p. 174). Physical anthropologists were less concerned with cultural heritage. Fields collected samples of bones, some of which were taken from bentwood boxes on mortuary poles, others dug up from graves to be used for the study of skeletal variation at several American Museums. The actions of these ethnographers has implications today. Graburn (2002, p. 19) writes of "the ethnographic tourist" and other researchers argue that modern tourists seek the unstaged and "totally Other" (Cohen, 2004, p. 360).

As Jenkins discussed, "primitive" crafts featured in the Modernist and Surrealist movements, which were also influenced by psychology and popular interest in the subconscious. Western artworks that reflect this trend include Picasso's "Les Demoiselles D'Avignon" (1907), Man Ray's "Noire et Blanche" (1926) and

Stravinsky's "Rite of Spring" (1913). It is not simply their rarity, but constructed perceptions of the pre-modern as authentically significant which give Haida arts and Haida Gwaii their magically real status. In 1908, Emily Carr started to document the Haida art landscapes, capturing war canoes and totems in her paintings, repeatedly depicting the supernatural giantess (Rohlfsen-Udall, 2000, p. 36). Carr linked European modern styles such as post Cubism to Haida artworks, deriving abstract shapes for her compositions, reframing indigenous crafts and stylising the myth of place much as Georgia O'Keefe did with the New Mexico wilderness. The mystical, "primitive" narrative about Haida art and land seemed to be a myth set in stone.

Magically real places

This section reinvents ideas of magical reality from the normative literature context. It is important to further investigate how places can be reframed as "authentic" by the tourist industry and to extend our ideas of the authentication of places into the imagination. The tourism industry's promotion of Haida Gwaii involves a mix of liminal landscape, living crafts, ghost villages, heritage traditions and pre-modern artworks, all of which are highly evocative for visitors, signifying the emic authentic. The practice of tourism is highly ritualistic and has been described as a "sacred journey" (Graburn, 1989, p. 21). Encountering the Other appeals to cultural tourists who, when visiting indigenous sites, are often said to be seeking an authenticity of self (MacCannell, 1976, p. 589). Authors such as Rickly-Boyd (2012) argued that tourism involves the Western search for the authentic and auratic. Benjamin (1936) unpacked the theory of aura, which he related to qualities such as "cultic value", unobtainability, rarity and singularity in the age of digital and mechanical reproduction:

> The definition of aura as a unique manifestation of a remoteness, no matter how near it may be" . . . Remoteness is the opposite of propinquity. The essence of remoteness is that it cannot be approached. Indeed unapproachability is one of the chief qualities of the cultic image. By its very nature, it remains remote no matter how near. Any propinquity lent by its embodiment as matter does not impair the remoteness retained from its constituting a manifestation.
>
> (Benjamin, 1936, p. 39)

The discussion of aura articulates that the closer we try to bring a rare object by reproducing it—by photographing it or circulating images of it—the further its essence retreats. The marketing of British Columbia as *Super, Natural* (which explicitly includes references to "authentic experiences") accentuates the uniqueness of place, creating an object of desire. Authenticity has long been a pivotal aspect of tourism; Wang (1999) seminally divided tourism authenticity into different categories of "objective, constructed, experiential, existential and postmodern". While much of the tourism discussed in this study could be said to be experiential, socially constructed or even objectively measured, the work emphasises postmodern authenticity by exploring how *Super, Natural* British Columbia is presented as a postmodern "simulacrum" (Baudrillard, 1981) mixing historical references with myth to create a legendary place. Gao *et al.* (2012) investigated the simulacrum of Shangri La, which they argued to be a "phantasmal destination", based entirely on myth made into reality by the tourist industry. Vidon *et al.* (2018, pp. 62–64) have remarked that theories of postmodern authenticity and theories of simulacra can be applied to "hypernatural" wilderness environments, such as the Adirondack Park. As the paper will demonstrate, the legendary cultural heritage of Haida Gwaii is the embodiment of a simulacrum, the magic remote, untouchable.

Tourists authenticate places in ways which can be hot (emotive) or cool (cognitive) (Cohen and Cohen, 2012). Lovell and Bull (2017, p. 10) suggested that hot and cool authentication has a broader spectrum; it can freeze, warm, superheat and even "evaporate" into "magically real" experiences. The "magically real" authentication of places was said by Lovell and Bull to be dreamlike, as tourists perceived the otherworldly and extraordinary, drawing on the socially constructed "cultural imaginary" (Chronis *et al.*, 2012). Tourists are primed to expect the "elsewhere" of folklore and magic by the *Super, Natural* British Columbian branding.

The phrase "magic realism" has been defined as "the embodiment of the imagination in the everyday world" (Lovell and Griffin, 2018, p. 4). Lovell and Bull (2017) detailed examples of film location tourists imagining structures which are immaterial on the sites of film locations and Lovell and Griffin (2018) examined light installations as magically real phenomena, creating an envelope of light around a solid structure. The literary genre of magical reality is itself a contested concept, indicating the duality and fragmentation caused by colonialism, a mix between magical old and realist new. Slemon (1988, p. 9) indicates that

> the act of colonization, whatever its precise form, initiates a kind of double vision or "metaphysical clash" . . . within the colonial culture, a binary opposition within language that has its roots in the process of either transporting a language to a new land or imposing a foreign language on an indigenous population.

Magical realism presents a Euro-American view of the magical nature of indigenous beliefs and deep-seated traditions, which may be treated as spiritually functional by those societies. While magic realism originated as a Latin American genre, it has lateral application for all marginal voices. Leal (1995, p. 121) advocates that magical realism is an "attitude" and a postmodern mixing of genres, to be expected in the fluidity of world where the appropriation and fusions of cultural references has been accelerated. For example, Strecher (1999, pp. 267–269) posits:

> Others argue for a more politicized, but equally region-specific definition of magic realism as a post-colonial discourse that rejects traditional Euro-American emphasis on realism and positivism in favor of a worldview that permits the "magical" to coexist with the "real".

The lack of hierarchy between old and new suggested by this comment doesn't fully address the way in which the "old world" is designed to seem more meaningful than the new for consumption purposes, as everyday objects are elevated by their auratic status as art and tourist icons.

Bringing these ideas together, the use of magic realism techniques in tourist promotion involves the "imaginative authentication" of places by visitors, (Lovell, 2019), evoking a storied world which suits the more-than-human sites on Haida Gwaii. Larrington (2015, p. 7) argues that we long "to be told stories, tales which draw their energy from the places where we live or where we travel". Hemme (2005, p. 75) also suggests, "travellers have a need for narratives and imaginaries with which to individually enliven and animate landscapes and thus experience them afresh". The practice has been observed in tourists by Chronis (2012) who has discussed how tourists visiting museums are "story-builders" using a combination of the collective imagination and historical sources. Legendary spaces enable tourists to inscribe and reinscribe intangible tales onto material culture, encompassing Faris' (2004, p. 1) definition of magical realism, which combine "realism and the fantastic, so that the marvellous seems to grow organically within the ordinary, blurring the distinctions between them".

That the "marvellous" seems to be organic in some places is an important point; Haida Gwaii's more-than-human world indicates an essential intersubjectivity. In Haida Gwaii, nature has in effect taken over; it is a vision of a future beyond the Anthropocene, with forests thickly carpeted in moss, abandoned ghost towns and totem poles splintering as they return to the earth. Harrison and Rose (2010, p. 238) discussed how sites such as Uluru are viewed by Aboriginals as "sentient places" and Carr remarked how as she stumbled upon the carving of D'Sonoqua in the villages of Haida Gwaii, the "fixed stare bored into me as if the very life of the old cedar looked out, and it seemed like that the voice of the tree itself might have burst from that great round cavity." Schwenger (2006, p. 37) has discussed Lacan's argument that objects, so trees or totems, gaze back at us in a way which is "diffuse", "a sense that seeing happens, though without an identifiable source". MacGaffey (1998, p. 217) compares properties of art and magic, observing that "the preferred term in recent catalogues is 'power objects'". These sacred sites, or power places, continue to attract tourists, as do "purifying and redemptive" nature myths (Hennig, 2002). The romanticisation of Haida Gwaii is not a new phenomenon; Alice Henson Ernst wrote of the Pacific Northwest in 1939, emphasising its distinctive intersubjective connections between human, spirit and animal worlds (Jenkins, 2004, p. 128).

> On the Northwest Coast the lines between the three interlocking worlds lines between have always been very thinly drawn: the human world, that middle territory in which man dwelt insecurely, confused between the voices and visions from the terrifying superhuman territory above in the spirit world and that much more cheerful but still frightening sub-human terrain beneath: the animal world.

By describing tourism as *Super, Natural* and fantastical, the British Columbian Tourist Board is creating an auratic, mythic place. If the landscape of Haida Gwaii embodies magical reality, its art and crafts artefacts and production depict and epitomise spiritual beliefs, living traditions, resilience and supernatural animus, which is inextricable from the magically real sense of place. Haldrup (2017, p. 58) wrote of the "magical objects in everyday life" as exercising a kind of "modern magic", conjuring affective and powerful feelings. Haida crafts tap into Western concepts of the poetic; Bachelard (1994) articulated how receptiveness enables people to detect the "reverberations" of objects, detailing how "the distant past resounds with echoes". What is magical to one culture may be another's reality. It is important to note that mortuary poles, while reified, have a functional value for the Haida, recounting legends and stories, acting as mortuary vessels and commemorative devices. Discussing the relationship of Mexican architecture to magic realism, Burian (1997, pp. 50–51) remarked,

> As Alberto Perez-Gomez has eloquently pointed out, the magical qualities of Mexican architecture come from the "love in a rough object that has little to do with sophisticated theories" yet much to do with the earthbound tradition of making and the desire of the individual craftsperson to express and leave her or his mark on a place.

The "love in a rough object" and belief in pleasure of art and crafts made implicitly for their own sake is evident in Haida culture, as is their connotations of status as ceremonial trade items.

It is important to note that, as Kutzner *et al.* (2009, p. 99) observed, tourist encounters have social, cultural and economic consequences. "With the strangers' gaze invading the space of indigenous inhabitants, comes the potential for a myriad of cultural impacts from acculturation to commodification, exploitation to loss of identity". In contrast, tourist art has been disparaged as "airport art" by Graburn (1967, p. 28), a degeneration of "authentic" traditional arts and crafts, posing that the integrity of local art is replaced by staged authenticity to serve the tourist market. Cohen (1988) suggested that some tourists regard "commercialised objects" as authentic if they contain traditional designs and were handmade by local artists. Tourism is a traditional, ritualistic practice which confers meaning on objects, as is the art market. Hennig (2002, p. 178) described how "unconscious faith in art's magical powers continues to structure present-day viewing rituals, particularly in the significance accorded to the authenticity of a work of art". Errington (1998, p. 141) referred to the way in which "authenticity has been transferred from the object to the customer" and it could be argued that the art of the Haida, even in the form of souvenirs, embodies their resilience and beliefs.

Strategic magic realism

De-mystifying the *Super, Natural*, branding involves tracing its thread through different strategic documents. Haida Gwaii could be argued to be explicitly constructed by tourist sites as to commodify the concept of an authentically legendary space, with magic realism creating a dichotomy between the new world and old.

The village of Ninstints on an island off the west coast of Haida Gwaii is a World Heritage Site and is described by UNESCO (nd) as

> Remains of houses, together with carved mortuary and memorial poles, illustrate the Haida people's art and way of life. The site commemorates the living culture of the Haida people and their relationship to the land and sea, and offers a visual key to their oral traditions.

Any changes to the mortuary posts of SGang Gwaay are heavily negotiated. The World Heritage Site's authenticity statement includes the following wording:

> SGang Gwaay is unquestionably authentic in terms of its location and setting, forms and designs, materials and substances as well as spirit and feeling. The property is an authentic illustration of the evolving Haida culture, as can be seen in the relationships between the forms and designs of the art and structures at the property, and contemporary Haida art. The property continues to hold significant spiritual value for the Haida and is still used today.

Yet a key aspect of the "magic" is that S'Gang Gwaay is also a dystopian site and a site of memory and mourning, reflecting the decimation of a culture which continues to tap into "last of their kind" myths. Ruins are described as fragmentary and emotional, stirring the tourist to complete them in their mind's eye and the ghost towns of Haida Gwaii are allowed – it is an ideological statement not to restore them – to be reabsorbed by nature. The UNESCO authenticity statement (nd), describes how adjustments are made to the mortuary poles: "after consultation with chiefs and elders, in 1995 four poles were straightened and stabilized and in 1997 an additional pole was stabilized in an effort to prolong the period before they return naturally to the earth". The action indicates that full restoration is not acceptable but a change can be made to stretch the timescale of their lifetime. Lovell and Bull (2017, p. 17) have discussed the state of "heritage flow"—the deeply emotional, almost vertiginous effect on tourists of viewing heritage structures which are warped and visibly affected by time. The attributes of the ruins, ghost town and the natural landscape of Haida Gwaii therefore feed the narrative of magicality, commodified by tourist producers at federal and regional levels and through the camera lens, incorporating myriad elements which can lead to the mystical romanticisation and mythologisation, forming a carapace over devastating events and geographies of absence of S'Gang Gwaay.

The magicality is thus compounded by the affective preservation of Haida Gwaii, yet it is described as a "living culture". The emphasis from federal to local levels on encounters with "living" heritage reinforce how "the needs of the community portraying its own cultural heritage and history come first; cultural tourism management is more effective than when it is imposed on a community by an outsider" (Agrusa and Albieri, 2011, p. 117). The Haida Heritage Centre, which opened in 2007, has a mission statement which also explicitly celebrates "living culture", reinforcing the underlying discourse of resilience and survival and avoiding Othering.

> Through the Haida Heritage Centre at Ḵay Llnagaay we celebrate the living culture of the Haida. Through our language, art and stories we share our relationship with the land and sea that which shapes, nourishes and sustains us. Ḵay Llnagaay protects and fosters Haida culture by reaffirming our traditions and beliefs, encouraging artistic expression, and serving as a keeper of all that we are. Ḵay Llnagaay is a place for the Haida voice to be heard. This is our gift to the world.
>
> (Haida Heritage Centre, nd)

Magic realism is the implicit focus of the federal and provincial strategies. Measures to stimulate experientially authentic tourism includes investment in the *Super, Natural British Columbia* brand, which is used "to strengthen brand appeal, increase urgency with travellers from key markets and draw industry partners toward greater alignment" making a "contribution of more than $4 million in funding to the Aboriginal Tourism Association of BC, spurring a doubling in the number of market-ready B.C. Aboriginal businesses to more than 300" (nd, p. 11). According to British Columbia's Tourism Strategy 2015–2018 "Gaining the Edge," the market share percentages of visitors to British Columbia are broken down as British Columbians – 50%, other Canadians (non-BC) – 21%, the US – 21%, Asia/Pacific – 4% and Europe – 3%. With investment comes pressure to mould reality to fit the brand and stage authenticity where it is required. The brand is designed to appeal to these markets by defamiliarising and accentuating the wilderness, imbuing it with magical attributes and accentuating the intangible cultural heritage to give it animus.

The *Super, Natural British Columbia* brand statement clearly promotes a magically real wilderness narrative to tourists, focusing on a simulacrum of the land: "A province shaped by nature. It has nurtured our people, our history, our culture . . . and our visitors . . . BC, like its people and its visitors, is 'wild at heart'". Vidon (2019) has described the "siren call of the wild", a statement about the compulsion of tourists to authenticate themselves in the wilderness, reminiscent of Emily Carr's (2009, p. 383) journal entry in 1937 discussing the role of women in art, referring to Canada as a "great rugged power that you are a part of". *The Guide to Indigenous Tourism* itself draws on and cultivates the more-than-human myth, stating, "Spend time with cultural ambassadors and nature enthusiasts – Haida people eager to share their venerable yet modern culture with you. Let them escort you into a magical world inhabited by whales, eagles, ravens and black bears". This is echoed in *The Guide to Aboriginal Tourism in Canada* (nd), which also stresses the magical reality of tourism, equating supernatural with wilderness:

> B.C. markets itself as "Super, Natural British Columbia," which is fitting. There's a pull towards the energies that live in our territories. We even have a name for supernatural in my language, in Kwa k'wala, which is "Nawala."

We have so much wilderness: old-growth trees, wolves, grizzlies, humpbacks and orcas. As diverse First Nations peoples, we are connected to these life forms. They are reflected in our art and our culture.

The Haida Gwaii Museum is enfolded in the supernatural simulacrum. According to its website, "the understanding of our inseparable relationship to the land, the sea and the supernatural – that which gives Haida Gwaii its incomparable natural and cultural character". These strategies and mission statements have outlined the blend of emphasis on living culture with the wilderness, wildlife and the supernatural connection of land and people to animals. While the iconic images of the mortuary poles are used on the front page of the Hello British Columbia and Go Haida Gwaii customer-facing websites, what is not mentioned as explicitly is the encounter with colonialism which led to their creation of the ghost villages; their memory has been diluted with different meanings.

Participatory planning

Further unpacking of the magical reality simulacrum is complicated by issues of ownership. As Niskala and Ridanpää (2016, p. 376) argue of the "exoticisation" of the Sámi culture in Lapland:

> In the studies of indigenous tourism one of the key arguments has been concerned with how ethnicity works as a resource for tourism promotion, which, in critical research, has been perceived as being part of the legacy from colonial times. . . . This concerns questions regarding for what purposes, how and on whose terms the cultures of indigenous peoples are represented in tourism promotion.

Local ownership and participatory planning appear to be cited throughout British Columbian strategic tourist documentation as indicators of authenticity. However, inclusion and collaboration does not always tell the true story; some voices can be louder, others silenced or othered. Tourism participatory planning includes the involvement of tourism businesses, community representatives and NGOs, both public and private sector and the local community for the approach to be sustainable (Timothy and Cevat, 2003, p. 181). The Federal First Nations Tourism Accord (1997) was created to establish a charter of respect and support between First Nations and tourism businesses and encourage participatory planning.

The tourism community is involved in the Haida Gwaii Destination Management Organisation (DMO) as partners and stakeholders; the GoHaidaGwaii brand, which launched in 2010; and the Haida Enterprise Corp., which owns Westcoast Resorts and other businesses, including a seafood operation. The Haida Gwaii Community Tourism Strategy indicates that place distinctiveness is grounded in local ownership: "The core elements of the Haida Gwaii/QCI experience that create its unique competitive position are its beautiful natural attributes combined with the living, vibrant Haida culture that welcomes visitors and offers authentic interaction and sharing" (nd, p. 5). As Hodgson (2007, p. 260) commented on Aboriginal tourists, "While there is no way of objectively measuring the authenticity of actors, one can understand where particular organisations fit within existing ideas about what kind of actors are more and less genuine". Thus, tourists can measure whether authenticity is staged through the signifiers of local ownership.

Many of the indigenous businesses likely to benefit from the investment are linked by the Indigenous Tourism Association of Canada (ITAC) that produces *The Guide to Indigenous Tourism in Canada*. ITAC specifies that if businesses are no less than 51% indigenous owned, they can participate in marketing and development opportunities and promotion, for example, inclusion in the *Indigenous Experiences Guide* (nd, p. 2). The organisation is designed to connect businesses and government, aiming to "support the growth of Indigenous tourism in Canada and address the demand for development and marketing of authentic Indigenous experiences". Both ITAC and the Haida Gwaii Community Tourism Strategy emphasise "authentic experiences", but fail to specify how an experience is designated to be "authentic". British Columbia's tourism strategy promotes "authentic experiences" which claim to offer "more meaningful and personal connections", (British Columbia Tourist Board, nd, p. 4) indicating that a backstage is revealed to tourists.

Living history is not simply a strategic approach, while a truly unstaged experience is hard to achieve between "host" and "guest" on organised, choreographed tours. Tourists search for unique insights and privileged access behind the scenes, witnessing demonstrations such as totem pole carving and participate in activities such as riding in a war canoe. Such "staged backstages" are established to create the illusion of unrestricted insights, yet they also keep visitors to a pathway, steering them away from fragile environments and have a part to play in the stewardship of true cultural privacy. They may also be transformative for those tourists involved. Local Haida guides, "the Watchmen", live on the islands of the archipelago. According to Hodgson (2007, p. 143), the background of tour guides is an important aspect of authentic tours. Describing her guide, she stated, "His function as an interpreter and educator of Aboriginal culture was seen as enhancing cross-cultural understanding and increasing the degree of genuine interaction between tourists and hosts". The Watchmen instruct visitors about life on the islands and explore the interpretative meanings of the World Heritage Site and iconic totem poles, communicating the magic and redressing the power balance through meaningful individual encounters.

Future meeting grounds

Postmodernist authenticity concerns hyper-commerciality, the equivalence of copies and originals and the acceptance of fakery and playful cross-referencing and representations of the Haida art and style are available at a "virtually real" level. The preponderance of copies can tip over into hyperreality, which dilutes the impact and aura of objects and, as suggested by Eco, 1986, p. 39), the copies may be better than, improve on and ultimately replace original artefacts.

For example, totem poles have been rescaled to a larger size in the Museum of Denver for performative effect; the cultural authority of the museum gives this staged effect creditability. Museum replicas can also replace repatriated objects and engender a dialogue between communities as part of object-orientated learning. Hyperreality is apparent in the work of Haida Manga by local artist Michael Nicholl Yahgulanaas, whose graphic novels have now taken shape in Vancouver in the form of a colossal raven Manga sculpture by Behdad Mahichi. Copies of iconic Haida design are adapted, appropriated, circulated and assimilated to an extent that affects both ownership and aura. Deloria (2004, p. 124) observes that representations of Aboriginal culture are ubiquitous; images of totem poles have also been rescaled to a tiny size on the back of juice containers, on clothes and as tattoos on the arm of the Prime Minister. While decolonialisation narratives are evident in the repatriation of remains, in the emphasis on authenticity and ownership on the round, images of Haida crafts have been digitally colonised and appropriated. Postmodernism could be said to dilute the affectiveness with its "magical hyperreality".

Ironically, perhaps the most experientially authentic experiences take place during staged performances of traditional dances, when costume, movement and sound are linked to traditional rituals which generate privileged, unstaged atmosphere, feeling of time-travel, communitas, performative authenticity and the inscribing of the past on the present (Lovell and Bull, 2017, p. **166**). The Raven Dance has been performed widely, and the distinctive costumes—which are accurately sourced and re-created—are ultimately folkloric, a style referenced by artists such as Kate Bush, whose "Fish People" are featured in her *Before the Dawn* performances.

Conclusion

The chapter has made an original contribution to research into authenticity by placing the *Super, Natural* branding of British Columbia in the context of magical realist theory. The findings imply that the everyday lives of one culture may seem fantastical to another when viewed through the tourist gaze.

The work highlighted how the objectifying ethnographic gaze froze Haida art as mystically primitive, pre-modern props for the Modernist era and fashion and objects of taste and curiosity. Haida Gwaii is a storied place, which is described as possessing "significant spiritual value" by UNESCO; as evoking experiential authenticity by the BC Tourist Board; and as resonant with the "energies of supernatural territory" by the Indigenous Tourism Association of Canada. Haida-owned local websites also promote their special relationship with land, spirits and animals; conveying folklore as part of functional, living traditions, magic realism is woven into the narrative. Having once been defined as pre-modern, the Haida now emphasise their contemporary "living culture" in local strategies that focus on the plural dialogue of "language, art, stories and voices" at the community hub of the Haida Heritage Centre. Participatory planning and ownership of the community is emphasised at federal and local levels as a key aspect of place brand authenticity.

Yet, the paper suggests that there is a continued strategic emphasis on the old world as a commodity of the new; although meaningful interpretative encounters, heritage stewardship and cultural performances to relatively small numbers of visitors to Haida Gwaii shift the power. In contrast, images of Haida culture digitally dispossess, appropriate and replicate, accentuating the multi-layered *Super, Natural* magically hyperreal simulacrum.

Recent narratives of decolonialism prompt a more ethically conscious approach to the stewardship of material culture, for example, acts of repatriation reaffirming the ownership of artefacts. In 2003, a mass repatriation took place of the remains of 150 individuals who were kept in the Field Museum of Natural History in Chicago. They were returned in carefully crafted traditional bentwood boxes accompanied by their descendants and reburied with ceremony and ritual. Yet, the memory boxes opened by the ethnographers cannot be reclosed. Visitors continue to gaze on the living culture and artists, the original, disintegrating longhouses and mortuary poles in place on the island of SGang Gwaay. As they do, magical realities shift between a landscape overlaid with myth, otherworldly nature, the memory of the horrific devastation of a culture and evidence of cultural resilience. For visitors, the *Super, Natural* will always be an untouchable, ever-retreating simulacrum.

Bibliography

Aboriginal Tourism Association of Canada (ATAC) 'Guide to Aboriginal Tourism 2016–2017'. Available at: https://aboriginalcanada.ca/corporate/aboriginal-tourism-experience-guide/ [Accessed 30 Jun. 2017].

Agrusa, J. and Albieri, G. (2011) 'Community Empowered Tourism Development: A Case Study', in Laws, E. (ed.) *Tourist Destination Governance: Practice Theory and Issues*. Wallingford: CABI, pp. 117–133.

Asplet, M. and Cooper, M. (2000) 'Cultural Designs in New Zealand Souvenir Clothing: The Question of Authenticity', *Tourism Management*, 21(3), pp. 307–312.

Bachelard, G. (1994) *The Poetics of Space*. Massachusetts: Beacon Press, 1994.

Barthes, R. (1957 and 2009) *Mythologies*. London: Vintage Books.

Baudrillard, J. (1981) *Selected Writings: Simulacra and Simulation*. Cambridge: Polity Press.

Benjamin, W. (1936) *The Work of Art in the Age of Mechanical Reproduction*. London: Penguin.

Berlo, J. C. (ed.) (1992) *The Early Years of Native American Art History: The Politics of Scholarship and Collecting*. Washington, DC: University of Washington Press.

British Columbia Tourist Board Strategy (2015–2018) 'Gaining the Edge'. Available at: http://www2.gov.bc.ca/assets/gov/tourism-and-immigration/tourism-industry-resources/gainingtheedge_2015-2018.pdf [Accessed 30 Jun. 2017].

Burian, E. R. (1997) *Modernity and the Architecture of Mexico*. Austin, TX: University of Texas Press.

Carr, E. (2009) *Hundreds and Thousands: The Journals of Emily Carr*. Vancouver: Douglas and MacIntyre Ltd.

Chronis, A. (2012) 'Tourists as Story-Builders: Narrative Construction at a Heritage Museum', *Journal of Travel and Tourism Marketing*, 29(5), pp. 444–459.

Chronis, A., Arnould, E. and Hampton, R. (2012) 'Gettysburg Re-imagined: The Role of Narrative Imagination in Consumption Experience', *Consumption Markets and Culture*, 15(3), pp. 261–286.

Cohen, E. (1988) 'Authenticity and Commoditization in Tourism', *Annals of Tourism Research*, 15(3), pp. 371–386.

Cohen, E. (2004) 'Contemporary Tourism-Trends and Challenges', in Williams, S. (ed.) *Tourism Critical Concepts in the Social Sciences*. Abingdon: Routledge, pp. 351–365.

Cohen, E and Cohen, S A (2012) 'Authentication: Hot and Cool', *Annals of Tourism Research*, Vol. 39, No 3, pp. 1295–1314.

Cooke, L. (2017) 'Carving "Turns" and Unsettling the Ground Under Our Feet (and Skis): A Reading of Sun Peaks Resort as a Settler Colonial Moral Terrain', *Tourist Studies*, 17(1), pp. 36–53.

Deloria, P. J. (2004) *Indians in Unexpected Places*. New Haven, CT: University Press of Kansas.

Dowie, M. (2017) *The Haida Gwaii Lesson: A Strategic Playbook for Indigenous Sovereignty*. San Francisco: Inkshares.

Eco, U. (1986) *Faith in Fakes*. London: Random House.

'Elginism'. Available at: www.elginism.com/similar-cases/the-fight-for-the-return-of-haida-remains/20030817/4452/ [Accessed 19 Feb. 2017].

Errington, S. (1998) *The Death of Authentic Primitive Art and Other Tales of Progress*. Oakland, California: University of California Press.

Faris, W. (2004) *Ordinary Enchantments: Magic Realism and the Remystification of Narrative*. Nashville, Tennessee: Vanderbilt University Press.

Gao, B. W., Zhang, H. and Decosta, P. L. E. (2012) 'Phantasmal Destination: A Post-Modernist Perspective', *Annals of Tourism Research*, 39(1), pp. 197–220.

'Go Haida Gwaii'. Available at: www.gohaidagwaii.ca [Accessed 29 Jun. 2017].

Goldberg, S. (n.d.) 'Race, Racism, History'. Available at: www.nationalgeographic.com/magazine/2018/04/from-the-editor-race-racism-history/ [Accessed 26 Apr. 2018].

Graburn, N. (1967) 'The Eskimos and "Airport Art"', *Society*, 4(10), pp. 28–34.

Graburn, N. (1989) 'Tourism: The Sacred Journey', in Smith, V. (ed.) *Hosts and Guests: The Anthropology of Tourism*. Philadelphia: University of Pennsylvania Press, pp. 21–36.

Graburn, N. (2002) 'The Ethnographic Tourist', in Dann, G. (ed.) *The Tourist as a Metaphor of the Social World*. Wallingford: CABI Publishing, pp. 19–40.

Grek-Martin, J. (2007) 'Vanishing the Haida: George Dawson's Ethnographic Vision and the Making of Settler Space on the Queen Charlotte Islands in the Late Nineteenth Century', *The Canadian Geographer/Le Géographe Canadien*, 51(3), pp. 373–398.

'Haida Gwaii Community Tourism Strategy 2009', p. 16. Available at: www.mieds.ca/images/uploads/HG%20QCI%20Complete%20Community%20Tourism%20Plan%202009%281%29.pdf [Accessed 10 Jun.].

Haida Heritage Centre. Available at: http://haidaheritagecentre.com [Accessed 30 Jun. 2017].

Haldrup, M. (2017) 'Souvenirs: Magical Objects in Everyday Life', *Emotion, Space and Society*, 22, pp. 52–60.

Haralambopoulos, N. and Pizam, A. (1996) 'Perceived Impacts of Tourism: The Case of Samos', *Annals of Tourism Research*, 23(3), pp. 503–526.

Harrison, R. and Rose, D. (2010) 'Intangible Heritage', in Benton, T. (ed.) *Understanding Heritage and Memory*. Manchester: Manchester University Press, pp. 238–276.

Hemme, D. (2005) 'Landscape, Fairies and Identity: Experience on the Backstage of the Fairy Tale Route', *Journal of Tourism and Cultural Change*, 3(2), pp. 71–87.

Hennig, C. (2002) 'Tourism: Enacting Modern Myths', in Dann, G. (ed.) *The Tourist as a Metaphor of the Social World*. Wallingford: CABI Publishing, pp. 169–187.

Hodgson, R. (2007) 'Perceptions of Authenticity: Aboriginal Cultural Tourism in the Northern Territory', p. 260. Available at: http://researchdirect.westernsydney.edu.au/islandora/object/uws:3806 [Accessed 28 Jun 2017].

Indigenous Tourism Association of Canada. Authentic Aboriginal Travel Worth 1.4bn Annually. Available at: https://indigenoustourism.ca/corporate/canadas-authentic-aboriginal-travel-experiences-worth-1-4-billion-annually/ [Accessed 29 Apr. 2018].

Jenkins. (2004) *Dream Catchers: How Mainstream America Discovered Native Spirituality*. Oxford: Oxford University Press.

Jonaitis, A. (1992) 'Franz Boas, John Swanton, and the New Haida Sculpture at the American Museum of Natural History', in Berlo, J. C. (ed.) *The Early Years of Native American Art History: The Politics of Scholarship and Collecting*. Seattle, Oregon: University of Washington Press, pp. 22–61.

Jordan, J. A. (2006) *Structures of Memory: Understanding Urban Change in Berlin and Beyond*. Redwood City, California: Stanford University Press.

Kutzner, D., Wright, P. and Stark, A. (2009) 'Identifying Tourists' Preferences for Aboriginal Tourism Product Features: Implications for a Northern First Nation in British Columbia', *Journal of Ecotourism*, 8(2),pp. 99–114.

Larrington, C. (2015) *The Land of the Green Man: A Journey Through the Supernatural Landscapes of the British Isles*. London: IB Tauris and Co. Ltd.

Leal, L. (1995) 'Magic Realism in Spanish America', in Lois, P. Z. and Faris, W. (eds.) *Magic Realism: Theory, History, Community*. Durham, North Carolina: Duke University Press, pp. 119–124.

Lovell, J. (2019) 'Fairytale Authenticity: Historic City Tourism, Harry Potter, Medievalism and the Magical Gaze', *Journal of Heritage Tourism*, pp. 1–18.

Lovell, J. and Bull, C. (2017) *Authentic and Inauthentic Places in Tourism: From Heritage Sites to Theme Parks*. Abingdon: Routledge.

Lovell, J. and Griffin, H. (2018) 'Fairy Tale Tourism: The Architectural Projection Mapping of Magically Real and Irreal Festival Lightscapes', *Journal of Policy Research in Tourism, Leisure and Events*, pp. 1–15.

MacCannell, D. (1973) 'Staged Authenticity: Arrangements of Social Space in Tourist Settings', *American Sociological Review*, 79, pp. 589–603.

MacCannell, D. (1976) *The Tourist: A New Theory of the Leisure Class*. New York: Solouker Books.

MacGaffey, W. (1998) 'Magic, or as We Usually Say, Art: A Framework for Comparing European and African Art', in Schildkrout, E. and Keim, C. A. (eds.) *The Scramble for Art in Central Africa*. Cambridge University Press, pp. 217–235.

National Aboriginal Tourism Research Project 2015 Canada. 'Economic Impact of Aboriginal Tourism in Canada'. Available at: https://indigenoustourism.ca/corporate/wp-content/uploads/2015/04/REPORT-ATAC-Ntl-Ab-Tsm-Research-2015-April-FINAL.pdf [Accessed 29 Apr. 2018].

Niskala, M. and Ridanpää, J. (2016) 'Ethnic Representations and Social Exclusion: Sáminess in Finnish Lapland Tourism Promotion', *Scandinavian Journal of Hospitality and Tourism*, 16(4), pp. 375–394.

Rickly-Boyd, J. M. (2012) 'Authenticity & Aura: A Benjaminian Approach to Tourism', *Annals of Tourism Research*, 39(1), pp. 269–289.

Schwenger, Peter (2006) *The Tears of Things: Melancholy and Physical Objects*. London: University of Minnesota Press.

Shields, R. (1991) *Places on the Margin: Alternative Geographies of Modernity*. International Library of Sociology. London: Routledge.

Slemon, S. (1988) 'Magic Realism as Post-Colonial Discourse', *Canadian Literature*, 116, pp. 9–24.

Strecher, M. C. (1999) 'Magical Realism and the Search for Identity in the Fiction of Murakami Haruki', *Journal of Japanese Studies*, pp. 267–269.

Talmazan Royal Couple Caught in Anti-LNG Protest on Trip to Haida Gwaii. Available at: http://globalnews.ca/news/2975504/royal-visit-2016-royal-couple-caught-in-no-lng-protest-on-trip-to-haida-gwaii/ [Accessed 19 Feb. 2017].

Timothy, D. J. (2011) *Cultural Heritage and Tourism: An Introduction*, Volume 4. Bristol: Channel View Publications.

Timothy, D. J. and Cevat, T. (2003) 'Appropriate Planning for Tourism in Destination Communities: Participation, Incremental Growth and Collaboration', in Singh, Shalini, Timothy, Dallen, J. and Dowling, R. K. (eds.) *Tourism in Destination Communities*. Wallingford: CABI, pp. 181–204.

Todd, Z. (2016) 'An Indigenous Feminist's Take on the Ontological Turn: "Ontology" Is Just Another Word for Colonialism', *Journal of Historical Sociology*, 29(1), pp. 4–22.

Udall, S. R. (2000) *Carr, O'Keeffe, Kahlo: Places of their Own*. Yale: Yale University Press.

UNESCO World Heritage Sites 'Sgang Gwaay'. Available at: http://whc.unesco.org/en/list/157 [Accessed 30 Jun. 2017].

Urry, J. and Larsen, J. (2011) *The Tourist Gaze 3.0*. London: Sage.

Vancouver International Airport 'At the Heart of the Airport'. Available at: www.yvr.ca/en/about-yvr/art/the-heart-of-the-airport [Accessed 3 Jul. 2017].

Vidon, E. S. (2019) 'Why Wilderness? Alienation, Authenticity, and Nature', *Tourist Studies*, 19(1), pp. 3–22.

Vidon, E. S., Rickly, J. M. and Knudsen, D. C. (2018) 'Wilderness State of Mind: Expanding Authenticity', *Annals of Tourism Research*, 73, pp. 62–70.

Wang, N. (1999) 'Rethinking Authenticity in Tourism Experience', *Annals of Tourism Research*, 26(2), pp. 349–370.

8 The authenticity paradox and the Western

Ken Fox

> The impossibility of rediscovering an absolute level of the real is of the same order as the impossibility of staging an illusion. Illusion is no longer possible, because the real is no longer possible.
>
> (Baudrillard, 1994, p. 19)

If, as Baudrillard suggests "illusion is no longer possible, because the real is no longer possible," why place the terms authenticity and Western genre in the same sentence? How can authenticity be claimed for the Western genre when in Baudrillard's (1994) terms it is its own simulacrum?

Despite this authenticity paradox, the Western genre continues to be a draw for film and television/streaming services' storytellers and directors. When America wants to tell stories about itself, it harness the Western's allegorical power. Since 2000 there has been a steady stream (pun only intended after 2008) of Westerns from Kevin Costner's return to the genre with *Open Range* (2003) after his Oscar-winning success with *Dances with Wolves* (1990). Terence Malick gave the origin story *The New World* (2005) the Hollywood art house treatment. High-profile director Quentin Tarantino mined the genre for *Django Unchained* (2012) and *The Hateful Eight* (2015). The Coen Brothers' love for the genre is evident in the Oscar-winning contemporary Western, *No Country for Old Men* (2007) and the remake of *True Grit* (2010). Filmmakers' ability to splice other genres with the Western can, at times, give generic hybridity a bad name, for example, *Cowboys & Aliens* (Favreau, 2011) and *A Million Ways to Die in the West* (MacFarlane, 2014). However, the genre's flexibility can help produce generic hybrids that challenge generic conventions, for example, the low-budget independent *Bone Tomahawk* (Zahler, 2015). Big-budget remakes of *3:10 to Yuma* (Mangold, 2007), *The Lone Ranger* (Verbinski, 2013), *The Magnificent Seven* (Fuqua, 2016), *Slow West* (Maclean, 2015), *The Revenant* (Inarritu, 2015) and *Hostiles* (Cooper, 2017) have shown Hollywood's willingness to get back in the saddle again.

Regarded by critics and scholars as revisionist Westerns, their claim to this title is often made around the notion of authenticity. In audience terms, this verisimilitude is claimed for marketing purposes as a marker of high-quality drama but in many cases, the touchstone for this authenticity is not necessarily historical research but a close approximation to the canonical Westerns of the past. These include *The Searchers* (Ford, 1956), *The Wild Bunch* (Peckinpah, 1969), *The Outlaw Josey Wales* (Eastwood, 1976) and *Unforgiven* (Eastwood, 1992). Westerns continue to be produced for television and streaming services, such as *Deadwood* (HBO, 2004–6), *Hatfield & McCoys* (Sony Pictures Television, 2012) *Frontier* (Netflix, 2016-), *Hell on Wheels* (AMC, 2011–16) *Westworld* (HBO, 2016-), *The Son* (AMC, 2017-) and *Godless* (Netflix, 2017). The ongoing production of Westerns for cinematic release including *The Sisters Brothers* (Audiard, 2018) and *The Ballad of Buster Scruggs* (Coen Brothers, 2018) demonstrate that the genre's allegorical power is not waning, and for the #Metoo and Black Lives Matter generation, the genre widely regarded as misogynistic and racist still has a role to play in national and international myth-making.

The Western's negative capability

To explain the genre's longevity and its ability to exist in a contradictory state, it appears there are at least two dimensions where audiences recognise the films are a mediated representation of a manufactured past, but their allegorical power still holds sway over the imagination. The Western genre appears to possess what poet John Keats (1817) described as "negative capability," existing in "uncertainties, mysteries, doubts without any irritable reaching after fact and reason." Westerns inhabit a creative/imaginative zone where the fake and the real co-mingle; where the structure of the genre disables the Native American gaze, obscures the female gaze and views of the world that are not white, male and dusty. Yet, filmmakers, writers, critics and audiences continue to use the term authentic in relation to the representation of the West in the Western genre. Rather than making an implausible judgment about how authentic a certain representation is, in this analysis there will be a focus on how authenticity is employed by filmmakers to embrace and hold at bay the overlapping states of fake West and true West.

Close analysis of *Hostiles* (2017) and *The Lone Ranger* (2013) explores how the Western's negative capability is chanelled through the cinematographic representation of the natural landscape drawing upon the discourse of the American sublime (Arensberg, 1986; Wilson, 1991). Artists such as Thomas Cole, Thomas Moran, Albert Bierstadt, Frederic Remington, George Catlin and photographers William Henry Jackson and Edward S. Curtis become the iconographic mapmakers for the development of the Western as a cinematic form (Buscombe, 1984, p. 1993). It is not only the sublime nature of the landscape represented by these artists that the Western embraces, it is also, I would argue, drawing on Riding & Llewelyn's (2013) discussion of Edmund Burke and the sublime "in Burkean terms the experience of pleasurable dread: the viewer staring into the abyss from position of comfort and safety." It is as if Burke (1775) in his treatise *A Philosophical Enquiry into the Origin of Our Ideas of the Sublime and Beautiful*, is describing cinematic spectatorship.

The invocation of the American sublime also appropriates Native American indigenous cultures/histories to represent the appearance of a more authentic connection to the land. O'Sullivan (2002, p. 77), reflecting on the exhibition of American Sublime at Tate Britain contends that:

> Defining the land – its vastness, its inhospitability, its savagery – was the first step towards legitimating its appropriation. Manifest Destiny determined that the territorial expansion of the US (which caused the displacement of Native Americans) was not only justifiable, but also divinely ordained.

This connection with a higher will, the conquering of the West as divine providence, is represented artistically as an experience of the sublime when viewing the wonders of the natural landscape. Yet, there is the strong sense that this natural landscape as represented is a social, political, ideological ad religious construction by the artists' hand (Arensberg, 1986).

The construction of a film story is a collaborative act with many players contributing to the overall achievement. Auteur theory (Astruc 1948; Truffaut, 1954; Sarris, 1962) identified the director as the most important element of moviemaking, while Mulvey's (1975) wonderfully dissenting voice drew attention to the male domination of the theory and its exponents. Without wanting to diminish any of the key roles in film storytelling, I want to make a case for the cinematographer or director of photographer (dop) as a co-equal. I argue that a key feature which helps tune audiences, critics, reviewers and academics in to the Western's negative capability—its ability to be false and true at the same time—is the depiction of the natural landscape. Masanobu Takyanagi, cinematographer on *Hostiles,* and *The Lone Ranger's* Bojan Brazelli harness the discourse of the American Sublime, and Remington's kinetic action pictures produce moving images that are not just serving the plot but creating a diegetic world where the reel and real West coalesce.

Cinematic landscapes in the Western

The relationship between how cinematic and television-constructed landscape is represented and the material landscape it depicts has become an established area of research in Film and Television Studies and in Cultural Geography. Film geographers such as Aitken and Zonn (1994) and Lukinbeal (2005) have drawn on and developed the work of film scholars, Wollen (1980), Higson (1984, 1987) and Lefevbre (2006) to set out a series of functions for the ways landscape might operate within television and cinematic representations. Lukinbeal (2005), drawing on the work of Jackson (1979) and Higson (1984), sets out how landscape functions in four ways in film: as space, as place, as spectacle and as metaphor. Applying some of these functions to the Western genre, and particularly to *The Lone Ranger* and *Hostiles*, enables a close analysis of how landscape through its use of cinematography manages to hold in tension the genre's uncertainties, claims of authenticity, binaries and competing ontologies. For the purposes of this analysis, I want to focus on landscape as place and metaphor as I argue these elements contain the conversion of cinematic space to place and work on the creation of landscape as spectacle through the cinematic apparatus.

Cinematic landscapes in *Hostiles*: putting the reel and the real in place

Hostiles, directed and written by Scott Cooper from a manuscript by Donald E. Westlake, follows the journey of veteran U.S. Cavalry captain Joe Blocker (Christian Bale). Blocker is ordered to escort one his most hated enemies, Cheyenne Chief Yellow Hawk (Wes Studi) and his family, from Fort Berrington, New Mexico Territory, to the Chief's home in the Valley of the Bears, Montana. The Chief is dying and at the insistence of the government, Blocker and a small team are tasked with making the 1,000-mile journey. Along the way, the group pick up a settler, Mrs Rosalee Quaid (Rosamund Pike), the only survivor of a massacre by Comanche raiders. Like Blocker and his second in command Master Sgt Thomas Metz (Rory Cochrane), Mrs Quaid is suffering post-traumatic stress. Named as melancholia by Metz in an early reflective scene with Blocker, the impact of the brutal wars with the Apache, Comanche and Cheyenne travel with the protagonists throughout. In the long tradition of journey narratives in Westerns from Ford's *The Searchers* (1956) to Mann's *Two Rode Together* (1954) and Eastwood's *The Outlaw Josey Wales* (1976), the central characters go through emotional and psychological developments reflected in part by their surroundings and how the landscape enables or disables their sense of agency.

Lukinbeal (2005) notes how the establishing or master shot in a film is a key element in the construction of landscape as place. Not only are the towns, forts and states named in graphics along the bottom of the screen but the extreme long shots of the varieties of landscape travelled through proclaims the importance of landscape as place in the film's narrative as well as providing a sense of spectacle for audiences.

The opening scene sets up the conflict of land ownership that is a recurrent theme in *Hostiles*. The establishing shot from a grassy hill shows a log cabin, a corral and stable. The camera moves forward and then cuts to the scene outside the cabin where a settler is sawing wood. The audience enter to the cabin where a woman is tutoring her children; the audience is alerted to the tension in this otherwise bucolic setting by the return of the camera to the distant hill where it now lingers. The film's soundscape signals the imminent sense of danger. The shape of horses and riders move past the camera to share the view the audience. A Comanche raiding party attack, murder the settler, Quaid, his three children, one a babe in arms, while his wife Rosalee Quaid manages to escape.

Settlers attacked by Native Americans—an often repeated element of the Western genre. When the opening credits roll and the title of the film is revealed, *Hostiles*, followed by a D.H. Lawrence quote: "The essential American soul is hard, isolate, stoic, and a killer. It has never yet melted." Apart from the quote, the audience appear to be in familiar genre territory. However, as the next scene unfolds we see Blocker and his men capture, torture and abuse an Apache runaway and his family. Brought back to the fort, the warrior is flung in to prison already overflowing with Apaches. In two scenes that sandwich the film's title, the director has confirmed the existence of racism, hostility, intolerance and brutal violence as part of the film's cinematic landscape. It is 1892 at Fort Berrington, New Mexico territory. The date is significant as it depicts a febrile atmosphere exists in this frontier territory just a year before Frederick Jackson Turner (1893) will declare the frontier closed in his treatise on the "Significance of the Frontier in American History."

At each stage of their journey, hounded by the band of marauding Comanche, lecherous prospectors and entitled white males who claim ownership of the land, the band of travellers dwindle, as the establishing shots hold for longer on the varieties of landscape. Rocky buttes and mesas as they set off from New Mexico territory, wooded uplands as they cross in to Colorado and verdant pasture as they enter Montana are all held for the audiences' attention, a Greek chorus reminding us the landscape is not simply a backdrop but essential to the progress of the narrative and the development of the characters.

A recurring note in the reviews of *Hostiles* is Takayanagi's widescreen cinematography. Philip Kemp (*Sight and Sound*, 2018) declares:

> The overwhelming glory of Hostiles, though, is Masanobu Takayanagi's widescreen cinematography. With New Mexico and Colorado depicting both themselves and adjacent states, he brings a rich, sweeping majesty to his landscape long shots, a lyrical sense of vast untamed spaces that recalls the classic era of the western. Time and again the distant prospect of a row of riders crossing a deep verdant valley or silhouetted against the skyline conjures up memories of Sam Peckinpah or Anthony Mann.

Mark Kermode in *The Observer* (Jan. 7, 2018) concurs:

> Japanese cinematographer Masanobu Takayanagi captures the harsh beauty of landscapes that shift from jagged rocks and perilous ravines to verdant valleys beneath glowing skies.

Landscape transforms from cinematic space in to cinematic place through the interaction of the characters with their surrounding but also by the way, as Kemp and Kermode suggest, canonical Western images are invoked to situate the audience in real-reel West. As the depleted party arrives at their final destination, The Valley of the Bears, they ride to the edge of a viewing point to take in the majesty of the setting. Straight out of the American sublime, the audiences' gaze is fixed on a version of home as authentic place for the Chief to breathe his last and become part of the land. This establishing shot appropriates the notion of a more authentic Native American link with the land but also throws the gaze back at the audience. As Rose (1993, p. 187) suggests, in relation to "a landscape's visuality", it "is seen as looking back, if you like, and having an effect in itself."

What an effect it has on our remaining protagonists. The tension of the incompatible mindsets of Indian killer, Joe Blocker, and sworn enemy, Yellow Hawk, are magically soothed by the journey but more importantly by the land itself.

Valley of the Bears, named on-screen to authenticate the place, but also miraculously, in terms of the vision supplied of the sublime landscape, to bring peace and understanding between previously implacable enemies. The work of the wilderness journey has been to bring them to a place of peace.

Arensberg's (1986, pp. 46–47) analysis of the American Sublime connects it directly to the ethos of Western expansionism and Manifest Destiny, and her critique offers a cautionary note for audiences faced with such a vision of natural beauty conjured by the cinematography in the Valley of the Bears scene.

Arensberg contends that the "actual source of this will" to the sublime,

> emanated neither from the natural landscape nor from some higher will but man's future power over the landscape expressed in the policy of western expansionism . . . Should anyone happen to feel guilty about this activity, the sublime, through its capacity to instill in the observer the concomitant feelings of diminution and awe, at once punishes the culprit in advance (through the feelings of being dwarfed by nature) and exonerates him through feelings of awe and rapture accompanying the vision of the sublime.

For all its good intentions, *Hostiles* remains trapped within the structures of the Western genre, a Sisyphean journey of advance and retreat. As Scott (Dec. 21, 2017) notes,

> *Hostiles* itself wants to be both a throwback and an advance, not so much a new kind of western as every possible kind – vintage, revisionist, elegiac, feminist. What makes the movie interesting is the sincerity and intelligence with which it pursues that ambition, heroically unaware that the mission is doomed from the start.

Scott (2017) draws attention to the film's power relations as an attempt to move towards tolerance. Cooper's liberal sensibilities are evident in an interview on the DVD extras where he refers to the divisions and intolerance created by the "make American great again" ideology and suggests his film as a counter to that divisive project. Blocker speaks to the Cheyenne in their native tongue, echoing Ethan Edwards (John Wayne) in *The Searchers*. He reads Julius Caesar in Latin, a civilized man and an Indian killer, who makes a profound journey in to the self. As with most Westerns that attempt a sympathetic portrayal of Native Americans we find in Blocker a character type that Prats (2002, p. 157) suggests at "once contains and denies the Indian . . . representing him while dispensing with him." Notwithstanding the intelligence and sincerity of Copper's film there is the overriding sense that the multi-dimensionality of Blocker's character is in stark contrast to the stoic, under-used and ultimately "noble savage" stereotype that surrounds Chief Yellow Hawk, despite a wonderful performance by Wes Studi.

As the closing credits roll against a backdrop of the map that shows the journey undertaken, with each change of credits comes a subtle movement across the map revisiting places and states covered on the journey. Robert Macfarlane (2018, pp. 97–98) identifies the difference between story maps and grid maps.

> A grid map places an abstract geometric meshwork upon a space, a meshwork within which any item or individual can be co-ordinated. The power of such maps is that the make it possible for any individual or object to be located with an abstract totality of space.

What the audience view as a background to the credits of *Hostiles* is clearly a grid map, but because we have made the journey with the characters it turns in to what Macfarlane names as a story map. The opening credits of *Game of Thrones* (HBO, 2012-) for example, manufactures this shift from grid map to story map beautifully and economically. According to Macfarlane,

> Story maps . . . represent a place as it perceived by an individual or a culture moving through it. They are records of specific journey, rather than describing a space within which journey might take place. They are organised around the passage of the traveller, and their perimeters are the perimeters of the sight or experience of that traveller.

In the case of *Hostiles*, the audience rewatch the film as part of their connection with the characters (travellers) and the abstract space of the map transforms in to place by the reiteration of the names of the places they/we have visited in the film's narrative. The slow zoom to the name Valley of the Bears as the final credits roll provides another layer of authenticity treatment to the overall project with the map, a metaphorical transformation of cinematic space in to three-dimensional place through our memory of the film.

Landscape as metaphor: the Tonto story

Wes Studi inevitably plays the supporting role in *Hostiles*, but *The Lone Ranger* epitomises the cognitive dissonance of Hollywood where a star actor, Johnny Depp, working with *his Pirates of the Caribbean* franchise director Gore Verbinski, undercuts the titular character, John Reid (Armie Hammer) and constructs the story of how the Lone Ranger got his mask through Tonto's narration. The film is Tonto's story.

The troubled nature of the film's production, the escalation in budget, the controversy over Depp's role as Tonto (although Depp claimed Native American ancestry on his maternal side, another Hollywood fantasy) and the lukewarm audience reception consigned the film to the arena of box office failure. The film suffers from an unevenness of tone; the director could not seem to determine whether it is slapstick or epic, fantasy or pastiche. Yet, there is much to confound in the film's placing of a Native American character (albeit, whitewashed) as narrator and ultimately the star of the film. The film is a perfect metaphor for the treatment of Native Americans in Hollywood representations: make a Native American the central character but ensure a white man plays the role.

If establishing shots are one of the key features of signalling landscape as place then the opening action scenes of *The Lone Ranger* are, in typical Verbinski style, excessive to the point of overpowering the narrative by the extraordinary vision of Western iconography represented. The town of Colby, Texas, 1869, awaits the arrival of the train as Verbinski borrows scenarios from *3:10 to Yuma* (Daves, 1957; Mangold, 2007), *Once Upon a Time in the West* (Leone, 1969) and *High Noon* (Zinnemann, 1952) to set up the film's narrative already overshadowed by the spectacular mesas and buttes in the background. The rapacious nature of American capitalism epitomised by the Western expansion of the railway, stolen mineral rights and the decimation of the Native American population, all strong elements in the narrative sits uneasily with the jokey tone. *The Lone Ranger* bears a Disney-fied relationship (more cognitive dissonance) to the social and economic critique of *Once Upon a Time in the West* and Cimino's *Heaven's Gate* (1980). A clear case of Hollywood having its ideological cake and eating it. By revelling in its fakeness, it could be argued that the film is more authentically inauthentic than *Hostiles* and yet they both draw from the same canon of Western iconography with similar visual but contrasting tonal effects.

The film begins in a Wild West Show in San Francisco, 1933, when a young boy wearing a Lone Ranger outfit walks in to see the exhibits and is confronted with "The Noble Savage" display where an ageing Native American plays the part. He comes to life, as it were, and recounts to the boy how the Lone Ranger got his mask. Johnny Depp channels Keith Richards playing Jack Sparrow as the elderly Tonto, with a dead bird headdress. Confronting the Native American stereotype head-on seems like a knowing Hollywood move but the irony is lost on the boy and becomes nothing more than a throwaway gag for the watchful viewer. Tonto's narration begins and the audience and the boy return to another Wild West set where the railroad snakes its way through a Monument Valley landscape. The director has taken every opening scene from a John Ford Western and amalgamated them in to a collage of made-up authenticity. This beautifully shot scene with the stunning backdrop clearly besmirched by iron lines and telegraph poles.

Lukinbeal (2005, p. 13) argues that "through the use of metaphor, meaning and ideology are appropriated in to landscape, the most common example of which is the attribution of human or social characteristics to landscape." In this scene's establishing shot it is not just meaning and ideology that are appropriated in to the landscape, it is as if the landscape is crying out for intervention. Handley and Lewis (2004, p. 6) suggest, "it is Nature or the natural landscape that serves as the sanctifying or authenticating stage of American Western experience," then *The Lone Ranger's* accidental or purposeful critique of rapacious capitalism and its impact on the landscape makes this film's metaphoric use of stereotypical Western iconography "hyperreal" (Eco, 1986). As Eco (1986, p. 8) has argued, "the American imagination demands the real thing and, to attain it, must fabricate the absolute fake." Brazerri's cinematography, in reaching for generic homage and influence, each landscape scene declares its absolute fakeness and in Baudrillardian terms as Felluga (2011) notes "there is no longer any distinction between reality and its representation; there is only the simulacrum." (Felluga, 2003, p. 86).

The film's readiness to celebrate fakeness is evident as the credits roll. Tonto, the ageing Wild West exhibit, vanishes from the exhibit case and is now seen walking towards the very mountains depicted in the canvas backdrop of the exhibit stage. The camera holds its position as Tonto, suitcase in hand, shuffles towards the distant ravine; a man returned to his natural habitat or another example of the vanishing American? When the metaphor becomes legend, print the metaphor.

Conclusion: in defence of the Western's true lies

The attenuated use of "negative capability" to conjure the Western's ability to be fake and still claim authenticity, to be false and true, to occupy a cinematic landscape that offers guilt and redemption, and as Arensberg suggests "diminution and awe" may in part explain the genre's longevity. Scott's (2017) review of *Hostiles* references Richard Slotkin's *Regeneration Through Violence* (1975), the first in Slotkin's extraordinary three-part dissection of America's turbulent and painful growth as a nation driven by westering expansionism and Manifest Destiny. According to Scott (2017), Slotkin argues, "appearances to the contrary the Western does not have fixed ideological meaning." Slotkin notes that the mythology of the frontier can be used "to reify our nostalgia for a falsely idealised past," or it can be used as a way of "imagining and speaking truth." Scott remarks how *Hostiles*, "like nearly every other good western and, for that matter, like the United States of America puts the genre in a state of contradiction with itself."

Once we begin to embrace the genre's incapacity for authenticity exemplified by two films, *Hostiles* and *The Lone Ranger*, with contrasting narrative intentions, box office results and critical acclaim, both drawing from the genre's bank of iconographic images, we can move beyond the rhetoric of authenticity. Both films deliver, in very different ways, the awe, wonder and spectacle of landscape as place and metaphor. The genre's conjuring trick, its negative capability channelled through the cinematic representation of the West's landscapes of spectacle and metaphor, its use of the American Sublime enables us to be more at ease with the genre's contradictory state. For, as Limerick (2003, p. 1) suggests, "the Real West and the Fake West end up tied together, virtually Siamese twins sharing the same circulatory system." While the marketers who promote Westerns might still use authenticity, the acceptance of the authenticity paradox, of the genre being in a state of contradiction with itself provides much more of an opportunity, as Slotkin suggests, for imagining and speaking truth.

Bibliography

Aitken, S. C. and Zonn, L. E. (eds.) (1994a). *Place, Power, Situation, and Spectacle: A Geography of Film*. Lanham, MD: Rowman & Littlefield Publishers, Inc.

Arensberg. M. (1986) (ed.) "Sublime Politics" in *The American Sublime*. Albany: State Univ. of New York Press. pp. 46–47.

Astruc, A. (1948) "The Birth of a New Avant-Garde: Le Caméra Stylo", in Graham, P. (1968) ed. *The New Wave: Critical Landmarks*. London: British Film Institute. pp. 17–23.

Barthes, R. (1972) *Mythologies*. London: Cape.

Baudrillard, J. (1994) *Simulacra and Simulation*. Translated by Sheila Faria Glaser. Ann Arbor: University of Michigan Press.

Baudry, J. (1974) 'Ideological Effects of the Basic Cinematographic Apparatus', *Film Quarterly*, 28(2), pp. 39–47.

Buscombe, E. (1984) 'Painting the Legend: Frederic Remington and the Western', *Cinema Journal*, 23(4), pp. 12–27.

Buscombe, E. (1986) 'The Idea of Genre in the American Cinema', in Grant, B. K. (ed.) *Film Genre Reader*. Austin, TX: University of Texas Press, pp. 11–25.

Buscombe, E. (ed.) (1993) *The BFI Companion to the Western*, New edition. London: British Film Institute.

Buscombe, E. (1998) 'Inventing Monument Valley: Nineteenth Century Landscape Photography and the Western Film', in Kitses, J. and Rickman, G. (eds.) *The Western Reader*. New York: Limelight Editions, pp. 115–130.

Buscombe, E. (2006) *100 Westerns*. London: British Film Institute.

Buscombe, E. and Pearson, R. E. (eds.) (1998) *Back in the Saddle Again: New Essays on the Western*. London: British Film Institute, pp. 77–95.

Cawelti, J. G. (1970) *The Six-Gun Mystique*. Bowling Green, OH: Bowling Green State University Popular Press.

Cawelti, J. G. (1999) *The Six-Gun Mystique Sequel*. Bowling Green, OH: Bowling Green State University Popular Press.

Cooper, S. (2018) *Hostiles*. 'DVD Extra, Making of Hostiles', Entertainment in Video.

Cresswell, T. and Dixon, D. (eds.) (2002) *Engaging Film: Geographies of Mobility and Identity*. Lanham, MD: Rowman & Littlefield Publishers, Inc.

Dr Ken Fox, School of Media, Art and Design, Canterbury Christ Church University.

Eco, Umberto (1986) *Faith in Fakes: Essays: Travels in Hyperreality*. Translated by William Weaver. London: Secker & Warburg.

Felluga, Dino. "Modules on Baudrillard: I: On Postmodernity" and "Modules on Baudrillard: II: On Simulation." Introductory Guide to Critical Theory. Purdue Univ. 31 Jan. 2011. (accessed June 12, 2018).

Handley, W.R. & Lewis, N (eds.) (2004) *True West: Authenticity and the America West*. Lincoln & London: Univ. of Nebraska Press.

Helphand, K. I. (1986) 'Landscape Films'", *Landscape Journal*, 5(1), pp. 1–8.

Hughes, T. (2016) 'Unpublished PhD Thesis: American Authenticity and the Modern Western 1962–84', Royal Holloway, University of London.

Jazairy, E. H. (2009) 'Cinematic Landscapes in Antonioni's L'Avventura', *Journal of Cultural Geography*, 26(3), pp. 349–367.

Kitses, J. (1969) *Horizons West: Anthony Mann, Budd Boetticher, Sam Peckinpah: Studies of Authorship Within the Western*. Bloomington, IN: Indiana University Press.

Kooijman, J. (2008) *Fabricating the Absolute Fake: American in Contemporary Pop Culture*. Jap Amsterdam: Amsterdam University Press.

Langford, Barry (2003) 'Revisiting the "Revisionist" Western', *Film & History*, 33(2), pp. 26–34.

Lefebvre, M. (2006) 'Between Setting and Landscape in the Cinema', in Lefebvre, M. (ed.) *Landscape and Film*. New York: Routledge, pp. 19–59.

Limerick, P. N. (1996) '"The Real West" in the Real West', commentary by Patricia Nelson Limerick, intro. Andrew E. Masich, Denver: Civic Center Cultural Complex. p. 13, quoted in Handley & Lewis (2004) *True West: Authenticity and the American West*. Lincoln & London: Univ. of Nebraska Press. p.1.

Llewellyn, N. & Riding, C. (2013) "British Art and the Sublime" in *The Art of the Sublime* Llewellyn & Riding (eds.) https://www.tate.org.uk/art/research-publications/the-sublime/christine-riding-and-nigel-llewellyn-british-art-and-the-sublime-r1109418, pp. 1–4. (accessed June 11, 2018).

Lukinbeal, C. (2005) 'Cinematic Landscapes', *Journal of Cultural Geography*, 23(1), pp. 3–22.

Lukinbeal, C. and Arreola, D. D. (2005) 'Engaging the Cinematic Landscape', *Journal of Cultural Geography*, 23(1), pp. 1–3.

Macfarlane, R. (2018) 'Off the Grid: Treasure Islands', in Lewis-Jones, H. (ed.) *The Writer's Map: An Atlas of Imaginary Lands*. London: Thames & Hudson, pp. 94–103.

Mulvey, L. (1975) "Visual pleasure and narrative cinema", *Screen*, 16(3), autumn.

O'Sullivan, N. (2002) 'American Sublime at Tate Britain', *Circa* (100), pp. 77–78. doi: 10.2307/25563823.

Prats, A. J. (2002) *Invisible Natives: Myth and Identity in the American Western*. Ithaca & London: Cornell University Press.

Rose, G. (2003) "Afterwords: Gazes, Glances and Shadows", in Robertson, I. & Richards, P. (eds.) *Studying Cultural Landscapes*. London: Arnold, pp. 165–169.

Sarris, A. (1962) "Notes on Auteur Theory", *Film Culture* 27, Winter 1962/63.

Scott, A.O (2017) "Review of Hostiles" in *New York Times* Dec 21, 2017.

Slotkin, R. (1973) *Regeneration Through Violence: The Mythology of the American Frontier, 1600–1860*. Norman, OK: University of Oklahoma Press.

Slotkin, R. (1985) *The Fatal Environment: The Myth of the Frontier in the Age of Industrialization, 1800–1890*. New York: Atheneum.

Slotkin, R. (1992) *Gunfighter Nation: The Myth of the Frontier in Twentieth-Century America*. New York: Atheneum.

Truffaut, F. (1954) "A certain tendency of the French cinema", *Cahiers du Cinéma* 31, January.

Tuska, J. (1985) *The American West in Film: Critical Approaches to the Western*. Westport, CT: Greenwood Press.

Wilson, R. (1991) *American Sublime: The genealogy of a poetic genre*. Wisconsin: Univ. of Wisconsin Press.

9 Playing at Westworld

Gunfighters and saloon girls at the Tombstone Helldorado Festival

Warwick Frost and Jennifer Laing

Introduction

The central premise of the popular television series *Westworld* (HBO, 2016–2018) is that modern-day tourists are able to find their true selves through immersing themselves in the American Wild West, as reproduced at the fictional theme park Westworld. In this high-tech fantasy world, small numbers of wealthy tourists engage in adventuresome scenarios similar to those commonly found in video games. Dressing and playing as Westerners, they engage in a conspicuous consumption of violence and sex. All components of the tourist experience are drawn from the iconic imagery of Western cinema. This includes costumes (one female tourist dresses in the poncho outfit favoured by Clint Eastwood in the 1960s), demeanour, plots and adversaries.

A central theme within this television series is authenticity. Such a concept is implicit in any consideration of media-driven theme parks, but in this instance, the discussion of authenticity is explicit. The story arc of one guest – Billy, played by Jimmi Simpson – constantly dwells on this issue. Billy is initially a reluctant tourist. He has been dragged along by his future brother-in-law Logan (Ben Barnes), partly because their family company is thinking of investing in the theme park. Billy is not very interested in playing at cowboys and is disturbed by Logan's excessive interest in sex and gunplay. In the early episodes, the impression is that Billy is passive and weak. However, as they move further and further into the game, Billy quickly changes, becoming violent and sadistic. Now it is Logan who is appalled, particularly when Billy repeats Logan's early words that Westworld is a place that allows one 'to be who you really are'. Later, having visited the park many times over the years, a now dissolute Billy explains that Westworld is 'a place you can sin in peace' as there are 'no consequences'. Billy has found his real self and it is deeply dark and disturbing.

Given that the technology of Westworld allows nearly any possible scenario to take place, it is surprising how narrow are the experiences depicted in the television series. Seemingly, all the paying guests want to be ultra-violent gunslingers. Their role models are the morally ambivalent anti-heroes of Western cinema, operating outside of the law and society. Drawing on cinematic imagery, they make limited choices. Nobody seems interested in playing the All-American hero of John Wayne. Rather than Henry Fonda's upright Wyatt Earp from *My Darling Clementine* (1946), the template of choice is Fonda's villainous Frank from *Once Upon a Time in the West* (1969). Nobody seems interested in peacefully immersing themselves within the Native American tribe, nor are any of the guests motivated by the pioneer myth embodied in Laura Ingalls Wilder's *Little House on the Prairie* (novels, 1870–1894; television series 1974–1983). The android 'hosts' that populate the park are racially diverse, yet the wealthy guests are apparently completely Anglo-Saxon. Authentically re-creating the West is purely for gratifying their sadistic and sexual urges.

If Westworld was real, would tourists actually behave this way? And, would they engage in existential musings about the nature of authenticity whilst immersed in a realistic fantasy world? We cannot research such questions in terms of a real Westworld, but there are American tourism experiences that partly contain some of its features. In particular, there are events in which participants are able to dress and play at being Westerners, engaging in behaviour quite different from what is normally acceptable in the real world and for which there are no consequences. In this chapter, we examine a case study of one such event, the annual Helldorado Festival at Tombstone, Arizona.

Helldorado festival, Tombstone

Helldorado originally began in 1929 to commemorate the fiftieth anniversary of this iconic Wild West mining town. The staging of the festival at that time benefitted from the combination of the growth of Hollywood films, increasing car usage and an elegiac nostalgia driven by the growing recognition that memories of the early days were diminishing. With the success of the initial commemoration, it was decided to stage Helldorado as an annual festival (see Frost and Laing, 2015a, 2015b for further background on Tombstone and the Helldorado Festival). The festival particularly focussed on the 1881 Gunfight at the OK Corral, involving a shootout between on one side Sheriff Wyatt Earp, his brothers and Doc Holliday and, on the other, the outlaw Cowboy Gang. As Tombstone evolved from a former mining town into a tourist destination, festivals like Helldorado complemented and activated its collection of shops, saloons and historical sites.

We undertook our fieldwork at the 2012 festival. Staged over a weekend, much of the festival program took place in the main street of Tombstone – Allen Street – which features an attractive streetscape of mainly nineteenth-century buildings. Of particular importance to this discussion was the ongoing series of reenactment 'skits'. These had four common components. First, they were staged in the middle of the main street. The main street accordingly became a stage, with spectators watching from both sides. It was also activated as a space in which the reenactors had 'returned' to the imagined past of the Wild West. Second, the skits told stories involving Tombstone in its Wild West period (c1880s–1890s). Interestingly, all of these stories were fictional. Whilst characters such as the Earp brothers were represented, there was no attempt to portray real-life historical episodes such as the Gunfight at the OK Corral. Third, the actors in these skits were enthusiastic amateurs, many of whom had travelled from other states. Interestingly, while groups paid a small fee to participate in the festival parade (a common practice in many history-themed festivals), there was no charge for performing a skit, indicating how highly their involvement was valued. Fourth, there was no narrator. The skits had titles in the festival program, but there was no attempt to explain their background or meaning. They simply occurred, as if the festival participants were back in the Tombstone of the 1880s.

In addition to the skits, large numbers of costumed participants simply paraded the streets of Tombstone. In undertaking such activities, they were aware that they were on show, being viewed and commented upon by others both in and out of costume. In essence they were 'promenading' (as people would have done in the Tombstone of the 1880s), engaging in conspicuous consumption and gaining status through their clothing, demeanour and behaviour. For those who participated in this way at Helldorado, this was perhaps also less demanding than the skits. They could visually adopt a persona, but they were not required to formally act it out.

Authenticity and re-enactment

Falassi (1987) argues that many events follow a *ritual structure*, comprising a series of common rites or components that have developed as a response to our basic human needs and beliefs. One of these is 'dramas', in which treasured stories are told or acted out. The skits at Helldorado fall into this category. In addition, two others of Falassi's rites are in evidence in some of these skits. These are 'conspicuous consumption' and 'reversal', and we will return to these later in this chapter.

Part of the appeal of these dramas is the perception of authenticity. This applies not only at Helldorado, but at many other reenactment events. In a study of a wide range of reenactors, Carnegie and McCabe found that they 'invested a huge amount of time, research, money and skills and enjoyment in developing a collection of material objects associated with authenticating the detail of their chosen period' (2008, p. 358). This emphasis on certain features of authenticity is a major component of many nineteenth-century reenactments in the USA. Two research studies are particularly worth noting.

Belk and Costa (1998) took part in thirteen reenactment events based on the Mountain Men of the Rockies in the period 1825–1840. They found that at these 'Rendezvous' events,

> participants socially construct and jointly fabricate a consumption enclave, where a fantasy time and place are created and experienced . . . The modern mountain man rendezvous as a fantastic consumption enclave is found to involve several key elements: participants' use of objects and actions to generate feelings of community involving a semimythical past, a concern for "authenticity" in recreating the past, and construction of a liminoid time and place in which carnivalesque adult play and rites of intensification and transformation can freely take place.

(1998, p. 219)

This psychological transformation took serious time and effort, particularly in a physical sense. Developing Mountain Men costumes and accoutrements was a slow and cumulative process. While gear could be purchased, part of the fantasy was engaging in ceremonial bartering as the Mountain Men did nearly two hundred years ago. Great respect and status was gained by those who were adept at making their own authentically styled costumes and equipment. Further status could be gained through exercising the appropriate braggadocio during the carnivalesque evenings, outshining others in prodigious displays of swearing, vernacular language, tale-telling and drinking (Belk and Costa, 1998).

Studying Civil War reenactors in the mid 1990s, Stanton found that at least half were male, White, middle class, conservative and military veterans (1997, p. 67). Many were confused and bitter with changes in society. Vietnam loomed large in these discussions, a military failure which they saw as indicating a lack of political leadership. As one informant told Stanton, 'The white male Anglo-Saxon of today really doesn't know what he's supposed to be'. He told her that he felt that many were drawn to reenacting 'as just looking for a simpler day, where they can go out and not feel guilty' (1997, p. 83). Stanton viewed this attitude as common to most of the reenactors she studied. She argued, 'these reenactors live in age when "immutable" truths are being challenged, and they use reenactment – either as a means of defence, escapism, or gradual accommodation – as a way of responding to those challenges' (Stanton, 1997, p. 77).

In events that feature dressing in costumes and reenactment, engaging in play may be important. This includes acting out a particular role or having fun in exaggerating certain features of the character or time period. This was certainly in evidence in the study of Mountain Men reenactments (Belk and Costa, 1998) and even in military reenactments (Elliott, 2007; Stanton, 1997). Of importance here is a study of play at the Parkes Elvis Festival in rural Australia (Jonson *et al.*, 2015). At this event, this study found that 'the attendees are active in creating the production – the play frame – whether it be dressing up, [or] posing with Elvises' (Jonson *et al.*, 2015, p. 488). A major feature of this festival was the wide variety of people dressed as Elvis and this worked well through 'the shared pretence that Elvis impersonators were "really" Elvis and that event attendees should treat them as such' (Jonson *et al.*, 2015, p. 491). This highlights that playing a character at a festival requires the participant to stay in character. If not, the result can be jarring and dissonant, as we observed at the Battle of Hastings reenactment in England, when the actor in the role of the Duke of Normandy started playing for laughs from the audience (Frost and Laing, 2013).

For those participating at Helldorado, popular cinema is an important form of information and the validation of authenticity. This highlights two important issues. The first is the perceived authenticity of costumes and weaponry. For Helldorado, the most influential film is *Tombstone* (1993), which was noted for the clothing of its main protagonists – particularly the Earp brothers, Doc Holliday and outlaw Johnny Ringo – who were dressed in rich and vibrant fabrics (Blake, 2007). This leads to an intriguing connection between the fictional skits and fictional cinema, with the amateur festival actors mimicking how many filmmakers focus on a convincing image in costumes and weapons rather than historical facts (Frost, 2006).

The second issue is the level of fictional invention in the skits at Helldorado. As noted earlier, none of the skits purport to be accurate history. Instead, they use Tombstone's history as background for their fictional tales. Again, this parallels the issues faced by filmmakers. The historian Rosenstone observed that Hollywood films about history typically took the view that 'as long as you get the look right, you may freely invent characters and incidents and do whatever you want to the past to make it more interesting' (1995, p. 60). Similarly, Seydor comments in his discussion of Director Sam Peckinpah's treatment of Billy the Kid:

> People who want history should read histories and biographies or watch documentaries. Novels, plays and films will always invent, distort or falsify because their priorities consist not in fidelity to facts but in telling good stories . . . adapting history to fiction, drama, or film always involves more elimination than inclusion, more reduction than expansion; composite characters have to be created; events discarded, changed, reordered, or invented for purposes of structure and plot.
>
> (2015, p. 278)

Usually, historical reenactments tend to follow the accepted historical story, requiring even amateurs to stick to the scripted plot. Examples of this include reenactments of the Battle of Little Bighorn (Elliot, 2007) and the Battle of Hastings (Frost and Laing, 2013), in which there is recognition that the audience knows what happened in reality and expect the reenactment to proceed accordingly. At Helldorado, however, such conventions are not adhered to. As with many Western films, the skits happily invent stories, with the only expectation that they conform to the shared mythology of participants and the audience.

The Western anti-hero

From the late 1940s onwards, Hollywood films focussed more and more on the 'anti-hero'. Rather than straightforward heroes, lead characters were more likely to be charismatic outlaws, or be flawed by obsessive behaviour. Key examples include characters played by Jimmy Stewart in *Winchester '73* (1950), Alan Ladd in *Shane* (1953) and John Wayne in *The Searchers* (1956). This trend was partly a reaction to the rise of television and partly through the application of *film noir* tropes, characters and actors to the Western (Slotkin, 1992). In the 1960s, the anti-hero reached a peak as the enigmatic and immoral stars of the Italian Spaghetti Westerns, best personified in the popular *The Good, the Bad and the Ugly* (1966). At Helldorado, the touchstone film is *Tombstone* (1993) and how it represents the Earp brothers and the Gunfight at the OK Corral. In that film, Wyatt Earp (Kurt Russell) is portrayed as a burnt-out violent ex-lawman, who simply comes to Tombstone to make money as a gambler. He is amused by the daily gunfights and uninterested in the growing influence of the Cowboys Gang. Wyatt only becomes involved to support his family, and when his brother is killed, he embarks on a violent vendetta. Accordingly, *Tombstone* – heavily utilised in how the participants dress and the advertising for the festival – is a tale of personal vengeance rather than justice.

All of this is played out in the streets of Tombstone during Helldorado. There are a few participants who opt for old-fashioned heroic role models. A close facsimile of John Wayne circa 1970 strides along and a number of skits feature sanitised versions of the Earp brothers. All of these are noticeably older participants. In contrast, there are more versions of Wyatt Earp drawing on Kurt Russell's portrayal. Particularly outstanding are a posing couple representing Russell's Wyatt and prostitute girlfriend Josephine (played in the film by Dana Delany). Strongly resembling the actors, they wear luxurious outfits and strike confident and aggressive poses. They do not, however, engage in any skits, rather preferring to pose for photographs.

In the skits, the anti-hero dominates and prevails. While those specifically playing the Earp brothers triumph, any other lawman is shot down in a hail of cussing and laughter. Villains stalk the street, overacting and enjoying comments from the crowd. Many of the skits are carnivalesque pantomimes in which evil wins. The more participants in each skit team, the greater the body count. The most elaborate sees 12 actors taking turns to die in ways suggestive of childhood games (Figure 9.1).

Figure 9.1 Westworld comes to Tombstone: reenacting a gunfight
Source: Warwick Frost

Amongst those promenading the streets, the most popular look is of a mean hombre. For some, this plays out as being costumed in the style of the Earps and Doc Holliday, with fancy suits balanced with prodigious moustaches and weaponry. In favouring this look, many are accompanied by women in saloon costumes (which will be discussed later). Others dress as dirt-stained desperados with outlaw personas. A number we talked to pointed out that they wear red bandanas around their waist, indicating membership of the infamous 'Cowboy' gang as depicted in the film *Tombstone*. Collectively, the ambivalent anti-hero lawmen and villainous outlaws greatly outnumber the straightforward 'good guys' of the mythical West. All of this suggests rituals of inversion within this festival (Falassi, 1987). Commemorating Tombstone's history, law and order is marginalised and law-breaking is celebrated.

Guns

The firing of rifles and pistols dominates Helldorado. Other weaponry is little in evidence, probably as firing a gun requires less skill and practice than using a knife or bow and arrow (it is noticeable that the tourists in *Westworld* fire guns for the same reason). The skits use blanks. Officials warn participants that they must only fire at each other from a safe distance – a blank charge can still injure – but this is occasionally ignored.

What is surprising for us is the prevalence of gun culture away from the performances. Two incidents illustrate this. The first arose from what started as a casual conversation. As Australians who regularly undertake research fieldwork in the USA, we find that brief chats with locals usually follow a standard path. These are typically very friendly, with Americans expressing interest and admiration of Australia and how they would like to visit one day. Our interaction that day at Tombstone was very different. Picking up that we were Australians, one person launched into a tirade against former Australian Prime Minister John Howard. Taken aback by his vitriol, we made our excuses and retreated.

The cause of this extraordinary encounter was John Howard's role in gun control in Australia. Following the 1996 mass murder of 35 people at Port Arthur, Howard championed the formation of the National Firearms Agreement, which limited ownership of semi-automatic weapons. The difficulty he faced was that it was the states – rather than the federal government – which had the constitutional power to regulate guns. It was only through threatening to stage a referendum to change the constitution that Howard was able to convince all the states to agree (Davies, 2016). Why then were these Australian domestic issues of interest in the USA? As we later learnt, America's National Rifle Association has featured John Howard in their advertising campaigns as an example of governments restricting gun use.

The second incident came the next day. Walking along a very crowded boardwalk in the main street, we observed that costumed outlaws with rifles on their shoulders and pistols in their holsters were in front of us. At this stage, it hit us that these were not replicas, but actual guns being carried through the crowd.

Somewhat later came the further realisation, that this being the USA and the Wild West, these were possibly loaded weapons. Whilst not personally confronting in the manner of the incident the day before, it was a striking realisation that the guns fired on stage were also present in the audience.

Saloon girls

Just as the film *Tombstone* provides the template for how men at Helldorado should act and look, it also provides a powerful model for many women. In the film, the morally ambivalent Wyatt Earp and Doc Holliday are matched with dynamic and attractive women in Josephine Marcus (Dana Delaney) and Big Nose Kate (Joanne Pacula). The characters and relationships portrayed in the film are essentially historically correct. Both women ran away from home as teenagers, headed west, worked as prostitutes and had long-term relationships with Wyatt and Doc. As shown in the film, Kate also probably had a relationship with Doc's adversary Johnny Ringo. Working in saloons, women like Josephine and Kate had financial independence, with Kate eventually owning a Tombstone saloon. Such narratives have often caused issues for how these women should be represented in the various Hollywood versions of the Wyatt Earp story (Blake, 2007; Frost and Laing, 2015b). In the film *Tombstone*, their backgrounds are explicit and there is no attempt at sanitisation.

At Helldorado, many women dress as saloon girls in the style of Josephine and Kate. This involves wearing luxurious and revealing clothing and adopting feisty characters. Whilst commonly seen amongst those promenading down the streets, these personas are also in evidence in the skits. Two of these skits are worth highlighting. In the first, two widows arrive in town. Finding that they have lost their savings, they wonder how they will survive. They are taken in by a female barkeeper (dressed as a man), who offers them a job. They become saloon girls, swapping their black mourning clothes for bright costumes. Their male customers – portrayed as hapless drunken miners – are quickly separated from their money and the widows embrace their new roles. In the second, Wyatt Earp and deputies arrive at a saloon seeking stolen money. The thieves are shot down in a gunfight, a common scenario in many of these skits. Then comes the twist. A trio of women come forth, challenging the lawmen. To the cheers of the audience, it is these women who then gun down Tombstone's erstwhile heroes.

What is striking with nearly all the women parading and performing as saloon girls is their ages. Many of them appear to be middle-aged. For these participants, adopting these personas provides them with the licence to act and dress in a way that is probably not normal for them. They can dress in revealing clothing and be sexually suggestive in conversation and behaviour. Our impression was that many of these women felt empowered through their performance, in the same way that the men were in playing at being gunfighters. There are similarities here with the Parkes Elvis Festival, in which researchers found that many older women dressed in the style of younger women of the 1950s and 1960s and concluded that 'the festival might have given middle aged and older attendees the licence to "let their hair down"' (Jonson *et al.*, 2015, p. 489). Intriguingly, the adventures of older female tourists are not featured in *Westworld* at all, while the female robotic hosts are all young and attractive. The older woman, in this television fantasy, is absent.

In contrast to the ultra-sexuality of the saloon girl characters, a small number of women dress in men's clothing. This phenomena was well-documented in histories of the West, with Calamity Jane the most well-known instance. Accordingly, one woman promenading the streets dresses as Calamity Jane, with short hair, trousers and buckskin clothing. Another, as noted above, plays a barkeeper in male clothing in one of the skits. What is completely missing is any representation of the attractive 'school marm' or innocent rancher's daughter as depicted in many Hollywood westerns (and indeed *Westworld*). Nor is anyone referencing *Little House on the Prairie*, with gingham dresses and cotton bonnets. It is interesting that despite the emphasis on pioneering women in the 'new history of the West' (Murdoch, 2001; Riley, 1999; White, 1991), women at Helldorado tend to nearly all embrace one persona.

Anglo-Saxon vision

The fictional theme park of *Westworld* is an Anglo-Saxon fantasy. All of the guests are wealthy and White. There are non-White actors in the cast, but they either play robotic hosts or technicians. Similarly, the representation of the West presented at Helldorado is almost exclusively Anglo-Saxon. As noted above in terms of women, this is despite much recent historical scholarship on the West attempting to give greater voice to the dispossessed and marginalised (Murdoch, 2001; White, 1991).

Amongst those parading the streets at Helldorado, the personas adopted seemed very White. We saw nobody dressed as a Chinese or Black and only one person in the guise of a Native American. What was intriguing, however, were the numbers who were drawing on the imagery of Josephine Marcus and Big Nose Kate from the film *Tombstone*. Both of these well-known historical characters were not Anglo-Saxon and it leaves it open to speculation as to whether their imitators were aware of this. Josephine Marcus had German-Jewish parents and Kate – whose real name was Maria Horony – was Hungarian.

Of the skits, two featured characters that were not Anglo-Saxon and this was a major part of the plot. One featured a Mexican who is thrown out of a saloon on the basis that 'they do not serve his type'. He returns with a pistol and shoots down all of his tormentors. Comic and carnivalesque, its contemporary message was well conveyed through its humour. The second featured two women performing a belly dance. At first, this seems incongruous in the setting of Tombstone, yet as they explained while performing, they were referencing an actual historical event. This was the 1881 performance of the 'Hoochie Coochie' dance (as belly dancing was then known), by 'Little Egypt' (Syrian-born Farida Spyropoulos) at Tombstone's Birdcage Theatre. Supposedly first performed in Tombstone, the dance was staged at various parts of the USA, including the 1893 Chicago World's Fair, over the next 20 years.

Conclusion

The Tombstone Helldorado Festival is a carnivalesque and entertaining melange. It does, however, raise important questions regarding the authenticity of such festivals featuring participants dressing up and acting as historical characters. As often noted, there is a tendency to dismiss such event behaviour as lacking seriousness and accordingly not worthy of academic study (Frost and Laing, 2013). Such a view is misguided, for the phenomenon exists and there are clearly complex meanings being worked out by those who participate. Such events might be fun, but they also have something to say about the society that creates them.

The reenactments at Tombstone place a great deal of emphasis on authenticity. This is, however, not authenticity in the sense that would be recognised in terms of cultural heritage studies. Whilst the participants invest a great deal of energy and resources in the conspicuous display of Wild West costumes, weaponry and other paraphernalia, there is almost no interest in re-creating historical stories. Instead, the mythology of the Wild West and Tombstone provides the background for enacting fictional stories in the main street skits. Furthermore, cinematic representations and tropes are important in crafting these stories. The result is that authenticity is perceived as critical to the spirit of the reenactments, rather than literal portrayal of a historically verifiable narrative. This view of authenticity is similar to those found at other historical reenactment events (Belk and Costa, 1998; Carnegie and McCabe, 2008; Frost and Laing, 2013; Stanton, 1997).

Part of the authenticity on display at Tombstone includes issues of violence, women and race. What is enacted at the festival is no longer acceptable in modern society. It may be that it is heavily influenced by cinematic representations of the mythological West. In providing a venue for the acting out of some anti-social behaviours, Helldorado is similar to the fictional Westworld theme park. Such play may allow participants to 'let off steam', expressing urges and behaviours that might otherwise be repressed. As Stanton (1997) observed, historical reenactments appeal to individuals who feel that changes in society have eroded their previous preeminence and status. Accordingly, they are seeking an authenticity through reenactment that allows them to temporarily return to what they see as a better and more satisfying past.

References

Belk, R. and Costa, J. A. (1998) 'The Mountain Man Myth: A Contemporary Consuming Fantasy', *The Journal of Consumer Research*, 25(3), pp. 218–240.

Blake, M. F. (2007) *Hollywood and the O.K. Corral*. Jefferson, NC: McFarland.

Carnegie, E. and McCabe, S. (2008) 'Re-enactment Events and Tourism: Meaning, Authenticity and Identity', *Current Issues in Tourism*, 11(4), pp. 349–368.

Davies, A. (2016) 'Port Arthur Anniversary: Seizing the Moment on Gun Control Changed Australia, Says John Howard', *The Sydney Morning Herald*, 26 Apr. Available at: www.smh.com.au/national/seizing-the-moment-on-gun-control-changed-australia-john-howard-20160424-godwg6.html [Accessed 23 Aug. 2018].

Elliott, M. (2007) *Custerology: The Enduring Legacy of the Indian Wars and George Armstrong Custer*. Chicago and London: University of Chicago Press.

Falassi, A. (1987) 'Festival: Definition and Morphology', in Falassi, A. (ed.) *Time Out of Time: Essays on the Festival*. Albuquerque: University of New Mexico Press, pp. 1–10.

Frost, W. (2006) '*Braveheart*-ed *Ned Kelly*: Historic Films, Heritage Tourism and Destination Image', *Tourism Management*, 27(2), pp. 247–254.

Frost, W. and Laing, J. (2013) *Commemorative Events: Memory, Identities, Conflict*. London and New York: Routledge.

Frost, W. and Laing, J. (2015a) *Imagining the American West Through Films and Travel*. London and New York: Routledge.

Frost, W. and Laing, J. (2015b) 'Gender, Subversion and Ritual: Helldorado Days, Tombstone, Arizona', in Laing, J. and Frost, W. (eds.) *Rituals and Traditional Events in a Modern World*. London and New York: Routledge, pp. 206–220.

Jonson, P., Small, J., Foley, C. and Schlenker, K. (2015) '"All Shook Up" at the Parkes Elvis Festival: The Role of Play at Events', *Event Management*, 19, pp. 479–493.

Murdoch, D. (2001) *The American West: The Invention of a Myth*. Reno and Las Vegas: University of Nevada Press.

Riley, G. (1999) *Women and Nature: Saving the "Wild" West*. Lincoln and London: University of Nebraska Press.

Rosenstone, R. A. (1995) *Visions of the Past: The Challenge of Film to Our Idea of History*. Cambridge: Harvard University Press.

Seydor, P. (2015) *The Authentic Death & Contentious Afterlife of Pat Garrett and Billy the Kid: The Untold Story of Peckinpah's Last Western Film*. Evanston, IL: Northwestern University Press.

Slotkin, R. (1992) *Gunfighter Nation: The Myth of the Frontier in Twentieth Century America*. New York: Athenaeum.

Stanton, C. (1997) 'Being the Elephant: The American Civil War Reenacted', Unpublished Masters of Arts Thesis, Vermont College of Norwich University, USA.

White, R. (1991) *"It's Your Misfortune and None of My Own": A New History of the American West*. Norman: University of Oklahoma Press.

10 Hidden in the mountains

Celebrating Swedish heritage in rural Pennsylvania

Katherine Burlingame

Introduction

> When morn with its splendor illumines the sky.
> Save where a star lingers to watch the night die.
> And the gray shrouding mist from the valley uprolled
> Is changed by the sun to an ocean of gold
> That bears on its bosom cloud land as fair
> As ever took shape in the realms of the air;
> Ah! who that, enraptured, has gazed on the scene
> Can forget the bright valleys and hills of McKean?
> – Jennie E. Groves, the *Reporter*, 1890
> (in Leeson, 1890, p. 58)

This poem about McKean County, Pennsylvania brings vivid images to my mind. I know what it's like to wake up at dawn and watch the thick morning mist slowly recede over the mountain tops. I know how the clouds swell into bulbous white cushions providing a backdrop for the endless towering mountains and deep rich valleys blanketed in dense, seemingly impassable forests. I know because I was born there, and my ancestors were born there. I don't merely know this land, I belong there. Part of my *self*, my identity, is deeply rooted in those mountains, those forests, and those small towns that freckle the landscape.

Over time, however, my sense of identity and belonging has been stretched and fragmented away from the small corner of my childhood. In an increasingly globalized and fast-paced world, identity has become more individualized and negotiable. Especially in the Western world, individuals play a significant role in shaping their own lives and thereby their identities beyond traditional categories of nationality, class, etc. (Paasi, 2003). As we become more connected with the world around us, however, we also grow more distant from what we know, and we often encounter a sense of rootlessness. Feeling detached from our ancestral homes or places of belonging, we are left to wonder who we are, where we come from, and where we truly belong. This crisis of belonging has come to the forefront of recent research that attempts to understand the effects of multiple and multiplying identities (Savage, 2008; Kramer, 2011; Bennett, 2018).

Nostalgia for the past not only resonates with individuals searching for rootedness, but also with rural communities whose unique histories become obscured with time. The once more isolated and homogenous communities of rural America, for example, now find themselves at a crossroads as younger generations move away or lack interest in their heritage, and with every passing year, fewer older residents still carry an authentic knowledge of the past. Recent research has therefore also shifted toward regional identities and rural communities (Paasi, 2003; Storey, 2010; May, 2013; Bennett, 2014, 2018; Csurgo and Megyesi, 2016) for their role in preserving their heritage and emphasizing local distinctiveness through, for example, heritage festivals.

This chapter examines the role of rural festivals in maintaining perceptions of authentic regional identity and preserving local history while fostering a sense of belonging and collective identity within associated communities. Drawing on an online survey and fieldwork conducted at the annual Mt. Jewett Swedish Festival in the state of Pennsylvania in August 2018, this chapter explores the complexity of identity in rural communities with unique histories as well as the increasingly insatiable need to belong and the pursuit of authentic experiences associated with rural America's immigrant past. Ultimately, this chapter highlights the role of heritage festivals beyond the preservation of heritage including fostering community bonds, emphasizing local distinctiveness, and preserving both authentic and inauthentic representations of life in rural America.

Identity and belonging

No one has simply one identity. Identity, rather, is a "complex, multi-faceted and many layered phenomena which refers to our understanding of both self and others" that is "multiple, fluid, unstable, and relational" (Storey, 2012, p. 11). Because people now play a much more active role in shaping who they are, identity is also often more individually than deterministically defined. However, an increasing spectrum of "optional identities" (see Lindholm, 2008, p. 53) has also led to a growing uncertainty of who we are as individuals and where and with whom we belong. Therefore, at the same time that our identities are more individually constructed, we nevertheless still crave a connection with others or with certain places in order to feel a sense of rootedness in an uncertain world. Increasingly, people tend to resist individualization and social isolation by renewing collective identities (Halbwachs, 1992; Castells, 1997; Inglehart and Baker, 2000; Jones, 2005; Anderson, 2006) that "are felt to be real, essential, and vital, providing participants with meaning, unity, and a surpassing sense of belonging" (Lindholm, 2008, p. 1).

Defined as "the process of creating a sense of identification with, or connection to, cultures, people, place and material objects" (May, 2013, p. 3), belonging is a much more fundamental process in our sense of self than rigid categories of identity. As May argues, identity categories are often very limited in their scope and miss out on the numerous feelings of connectedness or on the "complexity of being a person" where we are constantly negotiating different social categories as well as our "relationships with cultures, peoples and the material world" (2013, pp. 8–9). Using the term "belonging" also works more in the present day because it signifies the effort made to foster a connection or attachment with people and places beyond more traditional identity categories.

Through collective identities, people not only find a deeper meaning and sense of belonging within groups with similar cultural attributes, they also create attachments with places that foster connections with both intangible and tangible heritage. As Castells writes, beyond simplified categories of identity such as ethnicity or nationality, communities tend to develop based on "history, geography, language, and environment" (1997, p. 65). Belonging within a community, therefore, often coincides with a strong attachment to place, including the very land that keeps the community together. In fact, without place it is difficult or even impossible to foster a sense of belonging at all. As Miller (2003, p. 217) notes, the three main cornerstones of belonging are history, people, and place.

The question "where are you from?" in this sense is synonymous with "where do you belong?" "Home" is a place that has been charged with meaning, memory, and collective knowledge, and "to belong ontologically is to be implicated in a set of mutual obligations to care for the past and future of places and those who inhabit them" (Bennett, 2014, p. 670). Bennett refers to the collective knowledge of a place and its past as a gift that must be passed on between generations. This is similar to Bourdieu's concept of cultural capital (see Bourdieu and Nice, 2010) where collective identities are perpetuated through inherited traditions and customs. However, Bennett's study of families that have lived in one place for multiple generations also illuminates how a sense of belonging is "inherent in the daily actions undertaken by people who have inalienable connections to the places they inhabit" (2014, p. 669). Therefore, places with unique pasts and strong ancestral ties evoke not only a sense of belonging but also a sense of ownership and thereby responsibility to preserve and pass on knowledge. The history of a place is not merely a part of the past; it also plays an active role in maintaining the perceived authenticity of the community as well as the place.

The search for authenticity

In order to pass on cultural capital, communities re-create and represent the past through different narratives, memories, and material objects considered as genuine representations of the place and its people over time (Bennett, 2014). The need for authenticity emerges from the same desire to seek something real that helps us put down roots in a seemingly rootless world (Lindholm, 2008). As we face an unfamiliar and uncertain future, we seek familiarity within our own communities or those that ground us in the present and satisfy our nostalgia for the past. Authenticity is not only found within communities, however. It is also found in perceived authentic foods, music, and people. Therefore, it is important to distinguish these two layers of authenticity, both of which are tethered to a sense of belonging. In the first case, communities play an important role in maintaining perceived authentic elements of the past, giving members a sense of belonging and the responsibility for preserving those bonds. In the second case, belonging and authenticity coincide with experiencing such places as an "outsider."

As rural communities become more susceptible to the changes of the modern era, a nostalgia for the past becomes increasingly poignant. Maintaining a collective identity and authentic sense of place is vital for a continued sense of belonging, which ultimately binds together, gives stability, and ensures continuity for the community (Smith, 2006). One method employed by rural communities with unique pasts involves celebrating heritage through festivals, which "engenders strong emotions as collective memories and identities are either maintained and transmitted to younger generations or contested and remade" (Smith, 2006, p. 59). Communities emphasize what they deem important and authentic, which in turn attracts visitors who seek authentic experiences. By attending heritage festivals, visitors "construct a meaningful cosmos through the accumulation of multiplicity of authenticated sights (and sounds and tastes) that can be ordered like postcards in an album" (Lindholm, 2008, p. 41). Simply the *feeling* of engaging with the past becomes seemingly as authentic as actually *being* there.

The rise of the rural heritage festival

Since the late 1800s, Americans have had a growing interest in their genealogies – perhaps fueled by the unprecedented rise of immigrants in America referred to as "hyphenated Americans" (Kammen, 1991, p. 231) because of their multiple identities and having to deal with balancing their sense of belonging in their home countries with their pursuit of the American dream. Since a sense of rootedness and pride in their heritage formed an essential part of their identity, immigrant Americans (especially their descendants) played a large role in maintaining and thereby highlighting and performing distinctive traditions and customs. This interest in the lives of ordinary folk and their family lore, traditions, crafts, and products in the early 1900s was also due in part to the effects of industrialization and a growing nostalgia for threatened idyllic rural landscapes and the people living within them. Therefore, countless heritage societies were developed, and celebrations commemorating historical events, ethnic traditions, and cultures started to become more regular occasions.

However, it wasn't until the 1960s that a "lust for community" and a "turn to place" (Bauman, 1991, p. 246) reinvigorated a sense of curiosity for local heritage and what sets places apart. The process to determine "local distinctiveness" (Hawke, 2012) fueled a "heritage syndrome" (Kammen, 1991) or "ethnic renaissance" (Moberg, 1988) movement undertaken by hyphenated Americans in the 1970s. Communities were encouraged to appreciate their distinctiveness and to find ways to use their unique pasts to attract visitors and foster a sense of collective identity and belonging. Through the promotion of place, or "place branding" (Storey, 2010) in a heritage festival, for example, the local culture often becomes a resource, and the authenticity of the past becomes a staged performance where one's culture is put up for sale (Chhabra *et al.*, 2003; Storey, 2010). As Storey writes, with the commodification of place, "landscapes, local individuals or families, events, traditions, and building styles, along with other less tangible elements linked to culture and tradition, become resources pressed into the broader service of rural regeneration" (2010, p. 159). Therefore, the two levels of authenticity discussed before come into play with heritage festivals as community members reproduce their shared heritage and reinforce their sense of belonging, while visitors develop a sense of belonging or attachment to something they perceive as a "real" or "natural" part of the past because it has been legitimized by the participation of the local people with knowledge of the traditions (Chhabra *et al.*, 2003; Bennett, 2018). To find a rural heritage festival that maintains perceived authentic regional identities, preserves traditional heritage, and fosters a sense of belonging in the community and visitors alike, we must journey to a place hidden in the mountains.

Swedish heritage in rural Pennsylvania: the Mt. Jewett Swedish Festival

Between 1850 and 1925, one-fifth of Sweden's population, or 1.2 million people, moved to America. Though migration peaked in the 1880s, a steady flow of immigrants continued until after World War I. Looking for "an escape and a new beginning" (Barton, 1994, p. 33), many early immigrants were eager to assimilate to their new homes as quickly as possible. Barton writes: "Their emerging sense of identity reveals both the preservation of fundamental homeland values and the rejection of traditions and practices that had alienated them in Sweden" (1994, p. 4). With the "ethnic renaissance" in the 1970s, there was a huge boom in people's traveling back to Sweden as well as those learning about Sweden and going to work or study there. Around this time there was also a rise in Swedish and Scandinavian Festivals in areas whose populations had until recently been predominantly of Swedish origin. In 1970, one such festival began in the small town of Mt. Jewett, Pennsylvania and continues today.

High up in the rolling mountains encompassing the Allegheny National Forest in northwestern Pennsylvania in the United States lies McKean County. In the mid-1800s, Thomas L. Kane, a former Civil War colonel, received large amounts of land in McKean and Elk counties from stakes in a development company that had gone bankrupt after the war. Knowing the potential of the land, he advertised in Europe to attract Italian, German, and Welsh land purchasers. The largest majority to answer the call, however, were Swedish due to the land's close proximity to established Swedish farms, churches, and merchants, and also because of dire conditions in Sweden for members of the lower agrarian class, who were suffering from the effects of extreme weather on their crops and the oppressive nature of a rigid class system in the 19th century (Hulan, 1994). In 1865, the Swedish settlement of Kanesholm was established, and in 1870, Kane donated land for the construction of a Swedish Church – solidifying the Swedish presence in the area, which later dispersed into the newly built towns of Kane, Mt. Jewett, Smethport, and others. By 1925, McKean County contained the third-largest settlement of Swedish immigrants in America.

Today, Mt. Jewett is a small rural community with a population of 919. In 2000, 21% of the population claimed Swedish heritage (Pennsylvania State Data Center, 2000). With a median household income of only $32,500 and a decreasing population, Mt. Jewett relies on its annual Swedish Festival to boost the local economy, maintain its Swedish immigrant past, and encourage younger generations to take an interest in its history. Each year, different vendors and performers are invited to take part in the festivities, but there are a few consistent events including the competitions for King and Queen, the "maypole" dance, the meatball contest, and perhaps the most famous event, the "smorgasbord" offering traditional foods made from recipes passed down from first-generation immigrants.

Method

Two months before the festival, a short online anonymous survey was administered to three different groups on Facebook that are specific to the region and have very active members. The survey was only open to those who have attended the Mt. Jewett Swedish Festival. Of the 110 responses, many offered valuable insights into the meaning of the festival for the local community as well as the significance of preserving Swedish heritage in the region. Nearly 50% of respondents were over 50, and 7% of those were over 70, providing a variety of perspectives ranging from those who attended the festival when Swedish-speaking residents still existed to younger generations who experience a less-authentic festival and have different motivations for attending. Ninety percent of respondents said they were born in McKean County, ensuring the survey was directed toward the relevant community, and 72% claimed to have Swedish heritage, which they celebrate through holiday customs, flying the Swedish flag, traditional foods, language, and participating in the festival. To collect spontaneous impressions from a variety of participants, a random sampling of visitors and vendors were asked similar questions to those on the survey, and five informal interviews were conducted with festival organizers and long-term residents (eight people total as some were in pairs) ranging in age from mid-30s to early-80s. None of the interviews were recorded due to limited time and for the comfort of interviewees, but detailed notes were taken, and interviewees were asked to confirm any direct quotes. Survey and interview data were then coded, and four significant themes emerged that will guide the rest of the chapter's discussion: authenticity and local distinctiveness, embracing other cultures, heritage preservation, and community and belonging.

Authenticity and local distinctiveness

The Mt. Jewett Swedish Festival began in 1970, coinciding with an emerging fascination with local heritage and family history across America. Though one might assume the festival was started by Swedish immigrants or their descendants, multiple sources noted that two people of Italian descent were responsible. With a rise in heritage festivals across the country, they saw an opportunity to attract more visitors to the town and create an annual event to celebrate the town's history and reestablish the community's collective identity. Since Mt. Jewett consisted of predominantly Swedish families at the time, the festival was tethered to Swedish heritage. According to several sources, the festival would never have happened without the initiative of two ambitious community members, because "Swedes would never have done something like that for themselves" due to their more reserved and humble nature.

Also according to sources, there was a much stronger sense of authenticity in earlier years because the original festival involved many first- and second-generation Swedes. Over time, authentic elements have diminished. For example, the festival's "smorgasbord," an anglicized version of the Swedish *smörgåsbord* (a buffet with different hot and cold dishes), used to be a large event boasting different foods and versions of recipes from regions all over Sweden. Today, the table contains fewer traditional offerings because there are so few people left who still have authentic recipes or know how to make them. Similarly, over time the name and meaning of another one of the main events has been relatively confused. Outside of the building in which the smorgasbord is held stands a tall maypole decorated with flowers and greenery in the Swedish style. Despite the pole's authentic appearance, the organizers didn't know that the maypole in Sweden is known as the "midsummer pole" or that the celebration actually takes place during the summer solstice.

The festival also offered other events unrelated to Swedish heritage including carnival rides, a classic car show, and a parade with local companies, firetrucks, marching bands, and other participants (who sometimes waved a Swedish flag). A few vendors attempted to sell Swedish-inspired goods such as "traditional Viking knives" or textiles with phrases like "How Swede it is" or "But first, Fika." A shop called Kaffe Sol, open year-round, offers Swedish rye bread and *korv* (sausage).

With fewer *real* Swedes in the community, the festival has become decidedly less Swedish. An older resident noted that although some authentic elements are still present, the festival has been significantly watered down. Though he said there is still a Swedish "flare" and the desire to highlight the heritage, it is much more of a community event than a heritage festival. For example, he discussed how the King and Queen were once required to be fully Swedish. Now, anyone from the community can run, and even finding candidates can be challenging. Whereas contestants once wore traditional attire, today many costumes are inspired more by Disney characters than traditional Swedish dress.

An increasingly inauthentic festival also creates some tension with the remaining *authentic* Swedish residents of the town. For example, a man who came to Mt. Jewett directly from his farm in southern Sweden in the 1970s lamented how the earlier festivals were far more authentic. Still speaking with a Swedish accent, he said he doesn't attend because he experiences more authenticity at his home. His family celebrates Swedish holidays like St. Lucia, they continue to cook Swedish dishes including making their own *sill* (pickled herring) and *glögg* (mulled wine), and they sing songs in Swedish. They also used to hold a midsummer party to celebrate the summer solstice with other Swedish families in the area. His wife told me that in the past she had tried multiple times to inform the festival organizers that this was the proper use of the "maypole."

The Swedish heritage of Mt. Jewett nevertheless legitimizes the festival and the identity of the town and its residents. Having a rich history, one survey respondent wrote, "gives our little town a bigger meaning." Many other respondents noted that the festival should aim to maintain as much authenticity as possible because the town's Swedish heritage is one of the main reasons people come to visit.

Embracing other cultures

An interesting theme that emerged several times in the survey and in discussions with people at the festival related to how the festival, or growing up in Mt. Jewett, helped them embrace and appreciate other cultures. This theme is particularly important to discuss because it stands in stark contrast to many of the emerging contemporary stereotypes of white rural American communities.

Several survey respondents noted that the importance and benefit of preserving heritage goes beyond one's personal sense of belonging within a specific community. Instead, highlighting everyone's history is important in order to recognize the many different cultures that have shaped America over time. By celebrating different heritages, one argued, "we continue to add 'cultural flavor' to the melting pot that is America." Another added, we must always remain "aware that we are a country of immigrants." Similarly, there were multiple instances where respondents used phrases like "everyone's heritage is important" and "preserving any place's heritage is important" because it helps others to learn all the small pieces that make up the puzzle of America's rich immigrant past. One person noted, for example, that it wasn't only Swedish immigrants who made McKean County so beautiful, but that people from other countries shaped the landscape over time as well. Another wrote how growing up hearing Swedish spoken in Mt. Jewett helped increase awareness and appreciation for other languages. Several respondents also noted how all heritage is important because the desire to want to know what we are "made of" connects us all, and particularly in Mt. Jewett, even though not everyone in the town is Swedish, the festival provides the opportunity to "celebrate and learn from someone who could be different from you."

As mentioned before, many came to the festival simply out of an interest in the heritage. One woman whose father was born in Denmark was excited to hear about the festival not only because it makes her feel nostalgic for her Scandinavian roots, but simply because she "appreciate[s] human culture." Other visitors attended the Swedish festival as well as a nearby town's Italian festival at the same time, saying it was fun to do a "circuit of cultures" in such rural places.

Heritage preservation

Despite the fact that the festival celebrated its 48th anniversary, the role of the festival in preserving the town's heritage continues to decline. Several survey respondents noted the lack of participation, the loss of the older generation of

Swedes, and a lack of interest by younger generations as the main drivers threatening the preservation of heritage. The sense of urgency was expressed by many of the older respondents who said they have difficulty getting their children or grandchildren interested. One respondent wrote if the older generations "don't teach others, who will! It will be gone." Several younger survey respondents, however, expressed that they feel a significant amount of pressure to remain in the town or at least to take an interest in the preservation of its heritage by coming home and participating in the festival. That being said, there does seem to be some resilience in remaining residents of all ages who feel responsible for maintaining the festival and continuing to transmit traditions to younger generations. One person noted, "The town was built on Swedish heritage; it is our duty to continue with the Swedish traditions because we all need to have pride in our beginnings." Beyond the maypole dance, the Prince and Princess contest, and other events for children at the festival, there is also an effort to educate children in schools about the town's heritage. One respondent wrote that one school "had an assembly celebrating Swedish heritage in an effort to share with the kids the culture through food, dress, and traditions." Similar to others who expressed concern over a lack of interest in younger generations, the respondent also wrote, "If we don't find a way to interest the children in this celebration it will have no future."

In general, preserving the Swedish heritage helps to foster a sense of pride in the community, and as one person wrote, preserving the heritage "keeps the traditions alive – not just something we read in a book." There is common agreement that the loss of the heritage would have a negative impact on the community, which seems to indicate the festival will be held for years to come. As one respondent wrote, "The Festival is the heart of our town <3." Another wrote, "To put it bluntly we don't have a ton going for us. Less than 1000 people live here and it only seems to be dwindling. The Swedish festival is something that brings in a crowd every year and pumps a little bit of life into our town. It's sewn into the very fabric of our culture."

Community and belonging

Perhaps the most foundational and enduring theme that emerged from the survey and on-site research is the importance of the festival in maintaining a sense of community and belonging. By far the most common phrase was the universal sense of belonging during the festival for Swedes and non-Swedes alike. Although residents with Swedish heritage take pride in their ancestry, and although the festival will always be the *Swedish* festival, everyone gets to be Swedish for a day. One respondent wrote: "All in all, the festival brought such a sense of community and togetherness. . . . Of Swedish descent or not, we were all Swedes for those three days!" One of the festival organizers said despite the fact that "we don't have a lot of our Swedes [left], we just pretend we're all Swedish" during the festival. She noted, "we are all Swedish for the weekend." The same sentiment occurred frequently in the survey where residents of the town fostered their sense of belonging in the town by embodying the Swedish heritage. For example, one wrote, "I have no Swedish heritage but after living my life in Mt. Jewett, I am Swedish by heart."

"I may not be Swedish, but the heritage runs deep in Mt. Jewett and is something the residents of this small town are proud of and love to celebrate," said another. Perhaps the most powerful responses were those expressing the emotional sense of belonging that comes from living in the town. One person wrote, "I get choked up thinking about the town and the Swedish heritage. Though I have no Swedish blood, I feel that I am a Swede after growing up in Mt. Jewett. It is just a part of me. Some people don't understand this but it is a wonderful feeling to be a part of a small town."

Of course, for the residents who do have Swedish heritage, the festival is a nostalgic time for them to connect with their family's history. Several residents said they take pride in being one of the few remaining Swedish families in the town, and they feel a certain esteem knowing that their parents and grandparents played an active role in forming the community's identity. For these residents, perhaps the most historically significant tradition takes place every year during the festival at the octagonal Nebo church built in 1887 and modeled after the Ersta Kyrka in Stockholm. During the annual service, the remaining Swedish speakers give readings in Swedish – commemorating a time when every service in Mt. Jewett used to be in Swedish.

The success and continuation of the festival seems to go hand in hand with both the preservation of heritage as well as maintaining the town's unique identity. As one respondent wrote, "The festival allows for the opportunity to celebrate the real and 'imagined' community and folk traditions and culture that have served as the foundation for the community." While the Swedish heritage "serves as a bond for the community," the festival is the vessel through which those bonds continue to be forged.

The future of heritage festivals

Rural heritage festivals play a much larger role than merely celebrating heritage. The Mt. Jewett festival, and likely those in other rural towns, emphasize and preserve local distinctiveness while offering both authentic and inauthentic representations of the myriad cultural pasts of rural America. Perhaps most significantly, as attested by the Mt. Jewett festival, heritage festivals rejuvenate communities and strengthen familial bonds and friendships because they are the mechanism through which residents develop a sense of belonging and establish themselves as part of the town's collective identity. Though heritage festivals inevitably lose elements of authenticity over time, it seems that as long as people continue to come home and gather together, there will always be something worth celebrating.

Works cited

Anderson, B. (2006) *Imagined Communities*, Revised edition. London: Verso.
Barton, H. A. (1994) *A Folk Divided: Homeland Swedes and Swedish Americans, 1840–1940*. Carbondale: Southern Illinois University Press.
Bauman, Z. (1991) *Modernity and Ambivalence*. Cambridge: Polity Press.
Bennett, J. (2014) 'Gifted Places: The Inalienable Nature of Belonging in Place', *Environment and Planning D: Society and Space*, 32(4), pp. 658–671.

Bennett, J. (2018) 'Narrating Family Histories: Negotiating Identity and Belonging Through Tropes of Nostalgia and Authenticity', *Current Sociology*, 66(3), pp. 449–465.

Bourdieu, P. and Nice, R. (2010) *Distinction: A Social Critique of the Judgment of Taste*. Oxon: Routledge.

Castells, M. (1997) *The Power of Identity*. Oxford: Blackwell.

Chhabra, D., Healy, R. and Sills, E. (2003) 'Staged Authenticity and Heritage Tourism', *Annals of Tourism Research*, 30(3), pp. 702–719.

Csurgo, B. and Megyesi, B. (2016) 'The Role of Small Towns in Local Place Making', *European Countryside*, 8(4), pp. 427–443.

Halbwachs, M. (1992) *On Collective Memory*. Edited by L. A. Coser. Chicago: University of Chicago Press.

Hawke, S. K. (2012) 'Heritage and Sense of Place: Amplifying Local Voice and Co-Constructing Meaning', in Convery, I. and Corsane, G. (eds.) *Making Sense of Place: Multidisciplinary Perspectives*. Woodbridge: Boydell Press, pp. 235–245.

Hulan, R. H. (1994) *The Swedes in Pennsylvania, Peoples of Pennsylvania Pamphlet 5*. Harrisburg: The Pennsylvania Historical and Museum Commission.

Inglehart, R. and Baker, W. E. (2000) 'Modernization, Cultural Change, and the Persistence of Traditional Values', *American Sociological Review*, 65(1), p. 19.

Jones, O. (2005) 'An Emotional Ecology of Memory, Self and Landscape', in Bondi, L., Davidson, J. and Smith, M. (eds.) *Emotional Geographies*. Oxford: Ashgate, pp. 205–218.

Kammen, M. (1991) *Mystic Chords of Memory*. New York: Alfred A. Knopf.

Kramer, A. M. (2011) 'Kinship, Affinity and Connectedness: Exploring the Role of Genealogy in Personal Lives', *Sociology*, 45(3), pp. 379–395.

Leeson, M. A. (1890) *History of the Counties of McKean, Elk, and Forest, Pennsylvania*. Chicago: J. H. Beers & Co.

Lindholm, C. (2008) *Culture and Authenticity*. Oxford: Blackwell Publishing.

May, V. (2013) *Connecting Self to Society: Belonging in a Changing World*. Basingstoke: Palgrave Macmillan.

Miller, L. (2003) 'Belonging to Country – a Philosophical Anthropology', *Journal of Australian Studies*, 27(76), pp. 215–223.

Moberg, V. (1988) *The Unknown Swedes: A Book About Swedes and America, Past and Present*. Edited and translated by R. Mcknight. Carbondale: Southern Illinois University Press.

Paasi, A. (2003) 'Region and Place: Regional Identity in Question', *Progress in Human Geography*, 27(4), pp. 475–485.

Pennsylvania State Data Center (2000) *Mt. Jewett Borough 2000 Census Data*. Penn State Harrisburg. Available at: https://pasdc.hbg.psu.edu/sdc/pasdc_files/census2000/1604251632.pdf [Accessed 17 Aug. 2018].

Savage, M. (2008) 'Histories, Belongings, Communities', *International Journal of Social Research Methodology*, 11(2), pp. 151–162.

Smith, L. (2006) *Uses of Heritage*. New York: Routledge.

Storey, D. (2010) 'Partnerships, People, and Place: Lauding the Local in Rural Development', in Halseth, G., Markey, S. and Bruce, D. (eds.) *The Next Rural Economies: Constructing Rural Place in Global Economies*. Oxfordshire: CABI, pp. 155–165.

Storey, D. (2012) 'Land, Territory, and Identity', in Convery, I. and Corsane, G. (eds.) *Making Sense of Place: Multidisciplinary Perspectives*. Woodbridge: Boydell Press, pp. 11–22.

11 The triumph of trolls

The making, remaking and commercialization of heritage identity

Ann Smart Martin, Cortney Anderson-Kramer and Jared L. Schmidt

Introduction

A local historical society inaugurating a large new museum and history center is an uncommon occurrence in America today. But the sense of this community that they deserved their own physical museum and community space was generations in the making. The Driftless Historium and Mt. Horeb Area Historical Society in Mount Horeb, Wisconsin grew from a volunteer heritage organization that sought and managed a large number of records and heirlooms, many expressing the region's large and vibrant ethnic communities.

An interest in the formulation and performance of ethnic identity was deeply felt by many European immigrants and their descendants, but none more actively engaged in Mount Horeb than that of the Norwegians. Due to a combination of heavy immigrant settlement and the influence of strong personalities belonging to Norwegian descendants and enthusiasts alike, this town identity was realized through nearby heritage sites and articulated through multiple folklife expressions. Most importantly, however, the town had "branded" itself as a Norwegian place through successful heritage tourism, most notably a "Trollway" through the center of town.

This essay traces the two parallel tracks of museum and "Trollway" in the creation, formulation, and marketing of heritage identity and highlights the dramatic tension between real and imagined authenticity. It analyzes these processes by searching for the inter-tangled relationships of humans and the material world that created ideas of community and self. By using material culture methods, this project highlights the agency of the objects and landscapes themselves in changing the way people understand authenticity.

Welcome to the "Trollway"

More than thirty years ago, state officials planned to construct a bypass of Business Highway 18/151, abandoning the village of Mount Horeb. Rather than accept the fate of progress, the Mount Horeb Advancement Association, brought together by the Chamber of Commerce, drew on the town's legacy of Norwegian-American ethnic identity as a source of tourism (Gilmore, 2009; Lovoll, 1998).

Communities throughout the Midwest experienced a similar dilemma and some began incorporating their ethnic vernacular identity markers onto the landscape (Gradén, 2013). These vernacular features draw tourists to rural communities throughout the United States, sometimes influencing a rebrand as "concept towns" which function as a significant driving force in the local economy (Gradén, 2003).

The Midwest is home to a number of such communities that rely on their performed sense of ethnicity. Mount Horeb's neighbor to the east, Stoughton, similarly markets its Norwegian heritage through festivals like *Syttende Mai* (Norwegian Constitution Day) (Lovoll, 1998). To the south, New Glarus heavily advertises its Swiss identity through performances of the Wilhelm Tell play and the construction of chalet architecture (Hoelscher, 1998). Outside Wisconsin, Nordic ethnicity is promoted heavily in Lindsborg, Kansas, AKA "Little Sweden U.S.A." (Gradén, 2003) and the Danish community of Elk Horn, Iowa complete with windmill and Viking blacksmith shop (Larsen, 2013). Branding itself "The Troll Capital of the World," Mount Horeb joins communities engaged in transforming their vernacular landscapes, performing a sense of ethnic heritage through marketing the Norwegian folk creatures, trolls.

In popular culture, trolls are magical beings with deep resonance in Norwegian and later American culture that promise to either help or hinder humans. In

Figure 11.1 Trolls greeting visitors to the Open House Import Store on Main Street in Mt. Horeb. Wood.

Source: Image by Jared L. Schmidt.

J.R.R. Tolkien's Middle Earth and J.K. Rowling's world of witches and wizards, trolls are dimwitted, lumbering, and carnivorous beasts that live in the wilds. In other guises, they could be the tall-haired and gem-endowed dolls created by Danish woodworker Thomas Dam in 1959 and recently voiced over by musical pop star Justin Timberlake (2016). Children can imagine the behaviors of their own toy trolls, long-haired, squat, humanlike creatures produced by Hasbro, Inc.

The troll of Norwegian folklore similarly comes in many shapes, sizes, and variations, but is typically large and ugly with a penchant for violence towards Christians. Although threatening in size and strength, they are easily outwitted by humans (Christiansen, 1964). Mount Horeb's trolls, while similarly wanting in intelligence, play between these two extremes as they are a dramatically smaller, more humanoid variation with more amenable, albeit mischievous, natures.

Mount Horeb lies on the eastern edge of the Driftless Area, an unglaciated landscape stretching across southwest Wisconsin and portions of Minnesota, Iowa, and Illinois. Marked by rolling hills and river-cut valleys surrounded by prairie, it is easy to imagine a troll watching passersby from the shadows (Gilmore, 2012). Trolls in Norwegian lore prefer to make their homes hidden, often beneath mountains or in closed earthly features like caves. According to Janet C. Gilmore (2009), nearby caves like those at Blue Mounds State Park sparked Mount Horeb's community promoters to market these folk creatures as early as the 1960s.

While the physical landscape of the Driftless Area could captivate the imagination of a troll's habitat, the folk figures came from a cultural impact on the space. Although Norwegian immigration to Wisconsin began as a trickle in 1838, it developed into a flood which continues to shape the state's ethnic identity. By 1861, when Mount Horeb established its first Post Office, Wisconsin was home to approximately 44,000 Norwegians, increasing by another 14,000 a decade later, settling heavily in a 17-county area from Dane to Crawford Counties (Fapso, 2001; Lovoll, 2015; The Wisconsin Cartographer's Guild, 1998). According to the 1990 census, 35,449 Dane County residents self-identified as Norwegian, the highest in the state at 13.8 percent (Zaniewski and Rosen, 1998). As these immigrants settled down onto this landscape, so too did the trolls through oral narratives, folk beliefs, and folk art, becoming permanent residents of the Driftless.

Trolls also serve as narrative devices in Norwegian folklore to explain the existence of curious geographic features and are thereby known by the evidence of their landscape-shaping activities (Christiansen, 1964). The landscape peculiarities that inspire imaginative explanations entangle the mythological trolls within a very real geography, making these elusive and mischievous creatures into characters of fantasy and morality that have some bearing or footing in the shape of land. Because of the telling and retelling of stories based in place, the landscape comes to embody the trolls and their tales.

The trolls on Mount Horeb's "Trollway" are typically wooden sculptural folk creatures, often carved by chainsaw. Some are painted while others left natural, and mostly produced by artists in the community. The troll sculptures typically stand between three and five feet tall, making their smaller size seem childlike. They each display a unique personality, often satirically representing the business at which they are positioned. Like dolls too, they could be toys capturing our imagination and affection.

But the allusion is dashed through the artists' depictions of the body, hair, and clothes of adults from an unknown yesteryear. Like monsters or aliens, these trolls create the strong pull of something human, but a creature; friendly, but uncanny; grinning, but suspicious. Even the lion-like tail poking out from their clothes, resting along their feet with a domesticated contentedness, shares that ambiguity.

It is fitting, then, that the 37 trolls located along Mount Horeb's "Trollway" have themselves become vernacular landmarks, marking their new home as a heritage destination of marketed ethnicity. Barbara Kirshenblatt-Gimblett observes that, "To compete for tourists, a location must become a destination. To compete, destinations must be distinguishable, which is why the tourism industry requires the production of difference" (1998, p. 152). The "Trollway" is only the latest commercial version of a long line of difference-making activities signaling Norwegian heritage around Mount Horeb. One of the first critical links was the living history museum, Little Norway. From 1927–2012, the site presented an idealized version of 19th-century Norwegian life, combining vernacular architecture, costumed interpreters, and a replica of a *stavkirke* originally built in Norway for the 1893 Chicago Columbian Exposition. When Little Norway closed, local collecting museums, including the Historium, responded by acquiring some of the 7,000 individual objects housed in the *stavkirke*, thereby salvaging and memorializing the place (DuBois, 2018; Lovoll, 1998). In effect, the Historium became the heir to the Little Norway tradition, honoring its felt impact on the community and how it perceives itself as a part of a broader Norwegian-American place.

This inherited Norwegian community feeling is intrinsically reproduced by the Norwegian and Norwegian-American artifacts in Mount Horeb, especially through those painted with rosemaling. An iconic Norwegian adornment found throughout both the village of Mount Horeb and the Historical Society collection, rosemaling operates as a sign of the Norwegian community and identity, particularly in the Upper Midwest. Developed in 18th-century Norway, rosemaling features stylized two-dimensional floral motifs, geometric shapes, Rococo C-Stems, acanthus leaves, and an earth-tone color pallet. Immigrants transported rosemaling and its distinct regional variations to America, where it later became part of the mid-20th-century folk art revival (Gilmore, 2009). During the 1960s and 1970s, some Mount Horeb residents rosemaled all sorts of community surfaces, including the Open House Import Store, signifying their unified vision as a Norwegian-American community to themselves and an interested tourist public.

Simultaneously, efforts were undertaken by community leaders like Oljanna Cunneen to initiate annual performances of *Song of Norway*, a musical about Norwegian Composer Edvard Grieg (Gilmore, 2009). This public performance reinforced a direct unified sense of Norwegian identity in the community. Efforts such as these, and the annual "Scandihoovian Winter Festival," complete with snow-sculpture trolls, produces a sense that the community is authentically Norwegian. However, as Regina Bendix observes, authenticity, while connoting a sense of "original, genuine, or unaltered" is inherently a "plastic word," and that when invoked has become "increasingly intertwined with sociopolitical, aesthetic, and moral aspects with market concerns" (1992, p. 104).

Playing on the inherent plasticity of authenticity, Kirshenblatt-Gimblett suggests that heritage gives lifeways a "second life." Following her reasoning, Mount Horeb's evolution of how, why, and where ethnicity is acted out gives way to the idea of multiple lives, or distinctive historically rooted moments. The origin – arbitrary as it is – of Norwegian immigrants carrying with them ethnicity qualified only by their former residence in Norway followed by the second life performed through recollection and reiteration in a new country, and a third unfolded alongside heritage revivals of the mid-20th century. We might consider today's concept town a sort of fourth life. Leveraging generations of heritage performance rooted in the hills of the Driftless Area, the community has given heritage performance a new stage of life by establishing a museum alongside a city-wide campaign.

The Driftless Historium

Choosing to make boundaries that are geological rather than political, the Driftless Historium breaks with more typical local history museums and makes its mission broader, its history more diverse, and its artifacts and people more intimately tied to a place. Beginning with an explanation of the area's geologic processes, the permanent exhibition tracks a narrative between the Driftless Area and its peoples. The Indigenous Ho-Chunk people first worked the land and its bounty. In the first European migration came the farmers and lead miners mainly from Norway, Switzerland, England, Ireland, and Wales.

The objects in the exhibition are ample proof of the way that material culture can tell powerful and poignant stories of people and place, and amply address the museum's mission to collect, preserve, and make accessible the physical remnants of the Mount Horeb area's human and natural history "for the purpose of sharing the broader histories and prehistories of America through this localized lens" (Mount Horeb Area Historical Society, 2018).

The community's work to fundraise, build, and finish an 11,000-foot museum and history center in a short time was onerous. The staff and board's ultimate choice to complete the smaller changing exhibition space a year before the opening of the much larger exhibition's regional narrative was a pragmatic choice, partially based on the deep collection of European and European-American artifacts. These parts of the collection were also some of the most well-known and best-studied artifacts from folkloric study and fit an important trope of the town's and visitor's local interest. Nonetheless, the Historium's initial narration of the history of the Driftless Area based on the strength of its primarily Norwegian-American artifact collection forced the museum's education mission to work with and against the larger collection in order to provide an inclusive, broadly appealing experience.

In this way, Mount Horeb's organizing ambitions of regional and cultural commemoration and engagement exist in network with many other organizations that have likewise leveraged their cultural identities to serve their respective geographical and cultural communities. Building upon its emphasis on origins and development, the Driftless Historium also functions as a community center in which the local identity is crafted through curatorial authority, reinforced by the objects and subjects of inquiry, and performed through a rich programmatic calendar. The Driftless Historium follows a similar community- and research-oriented framework to Stoughton's Livsreise and Decorah, Iowa's Vesterheim in a celebration of Upper Midwest Norwegian and Norwegian-American heritage through exhibition. Furthermore, each of these institutions reinforces the strength of their collection and exhibition of material culture through activities, professional services, and performances. They provide genealogical services, educational presentations, festivals, heritage food-based events, folk classes, and general resources through community archives and libraries.

The Historium follows its roots in the Mount Horeb Historical Society that collected objects from the community, resulting in a collection that exists in a mutually reinforcing relationship with the communities' public ethnic identity. Once again, the sense of Norwegian community fostered a place for generations to continue to practice the traditions passed on from settler generations. In turn, the Historium examines and preserves this cultural memory while providing a venue for the community to learn more about their heritage, thereby strengthening that identifying sense of a Norwegian-American self within a Norwegian-American community. The Museum, the local historical society, early folklorists' study, and the "Trollway" hence offer competing versions of authenticity, but together reinforce a complicated expression of Norwegian-American experience.

An exhibition of heritage

Inaugurating the grand opening of its new public community center and smaller exhibition space, the Mount Horeb Area Historical Society drew upon its long partnership with folklorists and material culture scholars at the University of Wisconsin – Madison. As part of a museum course, faculty and students helped research objects, develop interpretation, and write a catalog and e-book for the exhibition. This cooperative venture fit the University's mission of partnering with local community groups to share and create knowledge, even earning a university award. Such a practice and its publicity add another layer of institutional authority and authenticity to a local museum.

The organizing goal of *Creators, Collectors and Communities: Making Ethnic Identity through Objects* was to communicate or illustrate representative moments of the local communities' ethnic identity as a continuous tradition while existing in constant dialogue with the American experience. Curators organized the exhibition into four distinct chronologically derived parts that they identified as representative of the Mount Horeb area ethnic lineage. The story is told through the phases "The Old World in the New," "Made in America with Foreign Parts," "Heritage Memorialized," and "Trolltown, USA."

When visitors enter the exhibition space, the first object they encounter is a Norwegian-immigrant trunk built by an unknown maker and painted by Ole Haugen circa 1835 (Figure 11.2).

Figure 11.2 Norwegian immigrant trunk. MHAHS 2014.073.0020. Pine case, iron hardware. The maker is unknown, but the trunk was painted by Ole Haugen, c. 1835. From the original Little Norway Collection acquired by Mt. Horeb Area Historical Society.

Source: Image courtesy of the Mount Horeb Area Historical Society, J. Bastien, photographer.

The Historium inherited the trunk from the Little Norway Collection, building its aura by association with local nostalgia and institutional authority. Just as the trunk crossed the Atlantic stuffed with wares for forging a new life in a new world, objects displayed in this first section evidentiate an unbroken tie between Mount Horeb and Norway. Produced in Norway, they are arguably authentic, linked to culture by time and place, legitimizing Mount Horeb's claim upon the Norwegian heritage.

Nonetheless, the trunk is still uniquely American through its ability to convey the American story of immigration. It is thus able to expand this specifically Norwegian object adorned with rosemaling into an object with broad American appeal. In preparation for their journey to the United States, emigrants carefully selected objects to survive an arduous voyage, build a new livelihood and home, and serve as memory markers from friends and loved ones (Dregni, 2011; Lovoll, 2015). In addition to packing enough food to undertake the long and perilous journey, emigrants filled their trunks with items ranging from clothes and raw materials to household goods and tools, some with deep sentimental and aesthetic value signified with memory. In the new world, the trunks themselves came to serve as emotional connections to Norway (Dregni, 2011; Gjerde, 1995).

The museum itself calls the question, "what gets saved?" Why and how did users and their families deem some objects worthy of careful preservation and memorialization and others not? Ultimately, what guiding principles did the museum use in the process of acquisition and curation? History museums across America exhibit trunks when telling the 19th-century immigrant story. They are integral to helping elucidate the physical pressures and sacrifices immigrants made on their journey, giving viewers a very tactile, relatable object upon which to project their historical imagination. But how did trunks come to be such important historical and cultural artifacts in museums (Maines and Glynn, 1993; Lowenthal, 1989)?

Trunks were common artifacts that retained their utility as household storage devices, even as they might be moved to attics or storerooms. They were thus less likely to be discarded. That overabundance led to their lessened value; even Little Norway's founder, Isak Dahle, excluded trunks from accession as late as 1928. Art historian and former curator of the Norwegian museum *Vesterheim* Marion John Nelson notes that immigrant trunks "were apparently also still too common to be considered in need of rescuing 'before it is too late'" (1994, p. 15). As these immigrant heirloom objects became increasingly rare, Nelson observes, their desirability to the museum world increased. When an object is deemed as worthy to display in a museum, the object's commercial and educational value is transformed. The trunks themselves hence refer to their original use as items of possession and memory and their later role as commodities or decorative museum objects. They demonstrate the multiple lives an object can have.

Through examples such as the Haugen trunk, we can view on a community-wide level anthropologist Igor Kopytoff's observation that "higher nongovernmental institutions or only quasi-governmental ones," such as museums and community organizations, play a vital role in the life of material culture, and that "who controls them and how says much about who controls the society's presentation of itself to itself" (1986, p. 81).

Curators chose objects for the first section that were aesthetically or functionally distinctive through the stories they elucidated about their originating culture. Alternatively, the second section, "Made in America with Foreign Parts," conveys the curators' interest in how immigrants and first-generation Americans re-created aspects of their heritage alongside American influence and new environmental limitations or opportunities. This part imagines how early immigrants adapted and expressed innovation through material culture within their new given environment. For instance, a Hardanger fiddle was a beautifully crafted Norwegian musical instrument but could be made in America from whatever non-traditional wood was available. Easily transportable as families moved west, Hardanger fiddles facilitated meeting new neighbors through the shared enjoyment of music, both old and new.

A central figure to this section, if not the entire exhibition, Aslak Olsen Lie was a regionally recognized furnituremaker whose cupboard in the Mount Horeb collection epitomizes the second theme (Figure 11.3). Lie was a well-established cabinetmaker in Norway and did not emigrate to America until he was 50 years old. Having already established his distinctly Norwegian style, as seen in the Skindrud cupboard with curvilinear ornamentations, Lie adopted notes of an American furniture aesthetic, specifically by producing his own version of the distinct stepback cupboard form. (The name is frequently used in the American antique market; the upper tier "stepped back" from the lower to create a small space to rest objects or foodstuffs.) Retaining the techniques that he learned from years of practice in Norway, Lie adjusted his style to meet American tastes. Crucially, though, he continued to introduce Norwegian elements, especially in painted ornament. Following the Norwegian tradition for painting furniture in bold colors, Lie colored his American form cupboard with a vibrant red exterior and contrasting blue interior visible through panes of glass.

While the rosemaled trunk evokes what was physically carried to America, immigrants brought their experiences, traditions, skills, beliefs, and so much more that was intangible, but evident in this section. None of these objects came directly from Norway nor originated from anyither nation. Makers created them through remembrance and adaptation to the needs and resources of their new home.

The third part of the exhibit, "Heritage Memorialized," perpetuates the sense of a natural, linear progression of ethnic representation in the village. Here objects produced by local folk artists turned tradition bearers draw upon their immigrant roots to transform found and raw materials into a celebration of ethnic revitalization. With a heavy emphasis on rosemaling and objects from Little Norway – 19 of 58 exhibited objects were from the Little Norway collection – this section unites art with artist, providing visitors an intimate glimpse into the creative process that transformed the community. As Bella Dicks (2000) and Laurajane Smith (2006) observe, the museum as institution serves as a site of heritage promotion reinforcing idealized conceptualizations of of local identity past, present, and into the future. This speaks to significant value community members place in preserving their past and performing their heritage, which the Historium enacts through material culture (see Gradén, 2013).

154　*Ann Smart Martin et al.*

Figure 11.3 Cupboard MHAHS 1984.011.0001. Pine, metal, glass. Made by Aslak Olson Lie c. 1860.

Source: Image courtesy of the Mount Horeb Area Historical Society, J. Bastien, photographer.

Wisconsin folk art represents a complex negotiation between the economic and geographical diversity, Old and New World influences, and technological innovation (Leary and Gilmore, 1987). Perhaps no individual personifies this negotiation of landscape and technology in Mount Horeb's folk art as much as Oljanna Cunneen (1923–1988). Born and raised in Dane County, Cunneen was instrumental in defining how Mount Horeb expressed its heritage. She excelled at rosemaling, but also created troll marionettes and textiles, and told Norwegian ethnic jokes. She personified Norwegian-American ethnicity for a generation of residents who spotted her around town sporting her *bunad*, a traditional Norwegian folk costume, and serving as a tour guide at Little Norway (Gilmore, 2009).

While the Historium exhibition features a wide variety of Cunneen's work, one piece demonstrates how she continuously engaged folk art in 20th-century contexts, stretching the potential of this 18th-century painting style. Folklorist Barre Toelken (1996, pp. 39–40) observes that traditions are subject to what he refers to as the "twin laws," in that they are both *conservative and dynamic* through their "retaining of certain information" while "compris[ing] all those elements that function to alter their features."

This negotiation of the twin laws is particularly pronounced in the A-line dress that Cunneen made for Mount Horeb resident Nancy Vogel in 1970 (Figure 11.4). Resembling the letter A in silhouette, these dresses are a practical easy-care form of fashion rooted in a sense of trendy practicality without excessive ornamentation, a feature that defined 1960s women's mass-produced fashion. This becomes a dynamic expression of traditional artistic form and local ethnic identity by blending folk art with mainstream fashion. Cunneen sewed this red woolen dress by hand and embroidered rosemaling patterns around the neckline and trim. The dress epitomizes Toelken's twin laws; providing a fixed moment in which an artist lent her skill to contribute to the transformation of a form and practice through and across centuries and mediums, all while assisting Vogel to feel more akin to her ethnic community.

The exhibition's final section playfully unites the village's ethnic identity with Norwegian folktales and the tourist-centered, heritage-driven economy by paying tribute to trolls. While these objects promote a larger sense of community engagement through self-promotion, some businesses hitched themselves financially to the "Trollway." Business owners began and continue to incorporate their company's theme or logo into the troll's appearance, even attracting fast-food franchisees to adopt the trollway theme. Persuaded by the novel attraction of the trolls, local businesses from bakeries to dentists draw upon the miniature creatures to drum up business and local affiliation. The exhibition's A&W troll sign (Figure 11.5) painted by Patricia "Pat" Edmundson represents a multilayered expression of Norwegian-American ethnicity as both franchise and artist transitioned outsider to insider.

Unlike her contemporary Oljanna Cunneen, Edmundson (1929–1993) is the descendant of English and Irish immigrants and did not claim Norwegian ancestry. Rather, she married and moved into the Norwegian-American community. Born and raised in nearby Janesville, Wisconsin, she and her husband, Wallace, moved to the Madison suburb of Monona in the 1940s. A naturally talented and largely self-taught painter, in 1968, at the height of the rosemaling folk art revival in the Upper Midwest (Gilmore, 2009), Edmundson enrolled in her first course out of a sense of artistic curiosity. She quickly demonstrated a strong proficiency, entering and winning area folk art competitions. According to their grandson Scott, Mount Horeb's "Norwegian charm" drew the Edmonsons to move there in 1974. He recalls that their four-bedroom home perpetually smelled of oil paints emanating from the converted rosemaling studio (Personal interview, April 19, 2016). When community leaders transformed Mount Horeb into a troll-centered concept town, she expanded her artistry by participating in the ethnic promotion of her adopted home.

Figure 11.4 Dress in a modern A-line cut featuring stitched rosemaling. MHAHS. 1990.014.0003 Made by Oljanna Cunneen c. late 1960s. Wool. Gift from its owner, Nancy Vogel.

Source: Image courtesy Mount Horeb Area Historical Society. J Bastien, photographer.

In 1984, Edmundson turned out a series of promotional paintings such as the A&W sign, demonstrating her knack for blending commerce with the village's plucky mascots (MHAHS 1999.098.0014). With a buck-toothed grin, this troll invites viewers to dine at this fast-food chain with its appeal rooted in a nostalgia

The triumph of trolls 157

Figure 11.5 A&W Root Beer and Restaurant company troll sign. MHAHS 1999.098.0014. Plywood. Painted by Patricia Edmundson, c. 1984.

Source: Image courtesy Mount Horeb Area Historical Society, J. Bastien, photographer.

for the drive-through era. Sporting an iconic orange sweater with the A&W logo on his chest, the figure validates the integration of this mass-marketed restaurant into the village. Engaged in a whimsical balancing act atop a hot dog, this troll effortlessly balances a glass of root beer in his right hand, a burger in his left, a salad on his heel and a vanilla ice cream cone at the tip of his right foot. Edmundson herself similarly balances the kitschy nature of Mount Horeb's conceptualization of trolls, navigating a space designed to be entertaining while drawing in passers-by off the highway to spend their money in the community.

Just as a strong preference for Norwegian objects made their way into the permanent exhibit, the curators of the inaugural exhibition pressed against the force of the collection's strong Norwegian presence in the effort to represent Mount Horeb's many ethnic communities. However, this tension was unavoidable because the Mount Horeb narrative ends in the present, a time in which Mount Horeb has tactically presented itself as Norwegian in character. In attempting to recognize the presence of a diverse pioneering community while emphasizing the shaping power Norwegians had on the town, the exhibition becomes a sort of fading away of diversity and an efflorescence of a particular culture. The exhibition's question then becomes, "How did Mount Horeb become the 'Trollway'?"

Concluding remarks

As suggested in this essay's title, the "triumph of trolls" lay in their capacity to help create a shared identity and to keep a village alive. Most cars still whisk by on a nearby highway. But Mount Horeb has successfully become a heritage destination where carved icons rooted in Scandinavian culture seem to wink at modern notions of authenticity. Their playful work in attracting tourism is real. The rosemaled town water tower and streetlamps beckon. Families find and track the carved sculptures scattered down a main street of shops and restaurants and share their experiences on social media. They might visit the museum to learn about the history of this place and about being an American in today's complicated era of immigration. While there, they might taste the themed troll ice cream or buy a vintage troll doll. As important, the Historium stands as museums always do. They use meaningful objects to create their authenticity, even as their authenticity creates meaningful objects. Trolls are magical things.

Works cited

Bendix, R. (1992) 'Diverging Paths in the Scientific Search for Authenticity', *Journal of Folklore Research*, 29(2), pp. 103–132.

Christiansen, R. T. (ed.) (1964) *Folktales of Norway*. Translated from Norwegian by P. S. Iversen. Chicago: The University of Chicago.

Dicks, B. (2000) *Heritage, Place and Community*. Cardiff: University of Wales Press.

Dreigni, E. (2011) *Vikings in the Attic: In Search of Nordic America*. Minneapolis, MN: University of Minnesota Press.

DuBois, T.A. (2018) 'The Migration of a Building: Representation, Replication, and Repatriation of an Emblem of Norwegian, Norwegian American, and Norwegian-American-Norwegian Identity,' *Scandinavian Studies*, 90(3), pp. 331–349.

Fapso, R. J. (2001) *Norwegians in Wisconsin*, Revised and Expanded edition. Madison: Wisconsin Historical Society Press.

Gilmore, J. C. (2009) 'Mount Horeb's Oljanna Venden Cunneen: A Norwegian-American Rosemaler "on the Edge"', *ARV Nordic Year of Folklore*, 65, pp. 25–48.

Gilmore, J. C. (2012) 'Restless Spirits on the Driftless Landscape', in Andrzejewski, A. V., Alanen, A. R. and Scarlett, S. F. (eds.) *From Mining to Farm Fields to Ethnic Communities: Buildings and Landscapes of Southwestern Wisconsin, Vernacular Architecture*

Forum Madison Meeting Tour Book. Madison: University of Wisconsin – Madison Dept. of Art History, pp. 34–55 and 335–341.

Gjerde, J. (1995) 'The Immigrant's Luggage: Observations Based on Written Sources', in Nelson, M. (ed.) *Norwegian Folk Art: The Migration of a Tradition*. New York: Abbeville Press, pp. 185–188.

Gradén, L. (2003) *On Parade: Making Heritage in Lindsborg, Kansas*. Rock Island: Upsaliensis.

Gradén, L. (2013) 'Performing Nordic Spaces in American Museums: Gift Exchange, Volunteerism and Curatorial Practice', in Aronsson, P. and Gradén, L. (eds.) *Performing Nordic Heritage: Everyday Practices and Institutional Culture*. Burlington: Ashgate Publishing Company, pp. 189–220.

Hoelscher, S. D. (1998) *Heritage on Stage: The Invention of Ethnic Place in America's Little Switzerland*. Madison: University of Wisconsin Press.

Kirshenblatt-Gimblett, B. (1998) *Destination Culture: Tourism, Museums, and Heritage*. Berkeley, CA: University of California Press.

Kopytoff, I. (1986) 'The Cultural Biography of Things: Commoditization as Process', in Arjan, A. (ed.) *The Social Life of Things: Commodities in Cultural Perspective*. Cambridge: Cambridge University Press, pp. 64–91.

Larsen, H. P. (2013). A Windmill and a Vikinghjem: The Importance of Visual Icons as Heritage Tropes among Danish-Americans. In: P. Aronsson and L. Gradén, eds. *Performing Nordic Heritage: Everyday Practices and Institutional Culture*, Burlington. VT: Ashgate, pp. 73–98.

Leary, J. P. (2006) 'Norwegian Communities', in Bronner, S. J. (ed.) *Encyclopedia of American Folklife*, Volume 3. Armonk: M. E. Sharpe, Inc., pp. 892–896.

Leary, J. P. and Gilmore, J. C. (1987) 'Cultural Forms, Personal Visions', in *From Hardanger to Harleys: A Survey of Wisconsin Folk Art*. Sheboygan: John Michael Kohler Arts Center, pp. 13–22.

Lovoll, O. S. (1998) *The Promise Fulfilled: A Portrait of Norwegian Americans Today*. Minneapolis: University of Minnesota Press.

Lovoll, O. S. (2015) *Across the Deep Blue Sea: The Saga of Early Norwegian Immigrants from Norway to American Through the Canadian Gateway*. St. Paul: Minnesota Historical Society Press.

Lowenthal, D. (1989) 'Pioneer Museums', in Leon, W. and Rosenzweig, R. (eds.) *History Museums in the United States*. Urbanna: University of Illinois Press, pp. 115–127.

Maines, R. P. and Glynn, J. J. (1993) 'Numinous Objects', *The Public Historian*, 15(1), pp. 9–25.

Martin, A. S., et al. (2017) *Creators, Collectors & Communities Making Ethnic Identity Through Objects*. Madison: L&S Learning Support Services. Available at: https://wisc.pb.unizin.org/mthoreb/.

Nelson, M. J. (1994) 'Material Culture and Ethnicity: Collecting and Preserving Norwegian Americana Before World War II', in Nelson, M. J. (ed.) *Material Culture and People's Art Among the Norwegians in America*. Northfield: The Norwegian-American Historical Association, pp. 3–72.

Rowling, J. K. (1998) *Harry Potter and the Sorcerer's Stone*. New York: Arthur A. Levine Books.

Smith, L. (2006) *Uses of Heritage*. London: Routledge.

Toelken, B. (1996) *The Dynamics of Folklore*, Revised and Expanded edition. Logan: Utah State University Press.

Townsend, Allie. 'All-Time 100 Greatest Toys'. Available at: http://content.time.com/time/specials/packages/article/0,28804,2049243_2048654_2049146,00.html [Accessed 19 Aug. 2018].

The Wisconsin Cartographers' Guild (1998) *Wisconsin's Past and Present: A Historical Atlas*. Madison: The University of Wisconsin Press.

Zaniewski, K. J. and Rosen, C. J. (1998) *The Atlas of Ethnic Diversity in Wisconsin*. Madison: The University of Wisconsin Press.

Interview cited

Edmundson, S. (2016) 'Non-Recorded Interview by Jared L. Schmidt', Madison, WI.

12 'It is yet too soon to write the history of the Revolution'

Fashioning the memory of Thomas Paine

Krysten E. Blackstone

Introduction

History remembers Thomas Paine for his radicalism. It has become, in many ways, his defining characteristic. Yet his radicalism is also the main controversy surrounding his memory. Once lauded as a patriot, subsequent generations have largely disregarded Paine in the American Revolution's standard mythology. This chapter considers the formulation of memory through studying the contested legacy of Thomas Paine in the United States during the periods where the popular memory of Paine was strong enough that individuals attempted to erect monuments. There are four monuments to Paine in the United States today. Groups erected each of these monuments during periods of increased memorialisation in the United States: 1839, 1881, 1952 and 1997. The monuments exist in three locations, each of which Paine resided in at some point during his time in America. In the nineteenth century, broad appropriation of Paine, even amongst otherwise conflicting factions, established the pattern of diversity and controversy over Paine's memory that subsequent generations followed. However, in the twentieth century, to become nationally acceptable, groups modified the memory of Paine to include only his accomplishments connected to the Revolution. This ignored his more controversial traits and made his memory more appropriate for popular consumption. The reformulation of Paine's memory between the nineteenth and twentieth centuries is indicative of the mollification of radical memory within America. Furthermore, it speaks to the popular understanding of the place of the Revolution in American culture and heritage.

Discussion

In the two centuries since the death of Thomas Paine, the debate over his place within American memory has been fraught and highly contested. Diverse swaths of society have evoked his memory in various capacities to accomplish their political agendas. A controversial figure in life, the debate surrounding Paine continued after his death. Born in England and integral to both the American and French Revolutions, Thomas Paine devoted much of his life to promoting his vision of radical change for the betterment of humanity. His four major writings help to explain his popularity, importance, and legacy. *Common Sense* garnered immense public support for independence during the American Revolution. *The Crisis Papers*, designed to foster morale and nationalistic sentiment, sustained the Revolutionary movement. Following America's victory, Paine returned to Europe and wrote *Rights of Man*, which encouraged revolution as a course of action when a government does not fulfil the natural rights of a people. Lastly, *Age of Reason* promoted religious reform, rather than social or political. Paine's writings focused on dismantling institutions and perceptions. Although Paine was crucial during the American Revolution, he was absent for the subsequent nation building, choosing instead to return to Europe and dismantle political systems there. As such, many in America have contested his importance. As individuals and groups appropriated his writings in any number of contexts, it is unsurprising that many have disputed Paine's legacy.

Fittingly, the memory of Paine in America is as varied as the man himself. Paine's designation as a radical coloured remembrance of him, and has influenced who chose to appropriate him and where the appropriations could be successful. To analyse his legacy within the United States, one only needs to consider the four periods in which the memory of Paine led to groups attempting to cement his memory by erecting or imagining monuments to him. Monuments successfully erected to commemorate Paine exist exclusively in locations where he resided: New Rochelle, New York, as well as Morristown and Bordentown, New Jersey. These towns erected the monuments to Paine during periods where memorialisation focused on the Revolution was rampant throughout the United States. In many ways, the erected monuments reflect more strongly on the towns themselves than they do Paine. The towns' ability to claim Paine simultaneously as a figure of the Revolution and their town is the key to the monuments' existence. Although the three different monuments and one bust reflect various aspects of Paine's life and emphasise different facets of his work, their existence relied on his domicile. Attempts to erect monuments in locations he did not live, even during the same periods as those standing today, were all unsuccessful.[1]

To understand a monument fully, several factors merit consideration. Understanding the society surrounding a monument and the individuals or groups that envisioned it is key to understanding the monument's creation. One must also consider the physical monument: the subject, design and rhetoric all contribute to the message it portrays. Moreover, the intended reception of the monument is crucial. Who is the intended audience? Has the reception of the monument changed throughout time? Underlying all of these factors is location. The message of the monument is often designed to correspond with its position. Placement of the monument affects the reception of it as it determines its audience. Where monuments were successfully and unsuccessfully erected is revealing when considering how groups and individuals have chosen to commemorate Paine in the United States (Lowen, 1999, p. 459). Significantly, all of these factors contribute to the authenticity of the statues. Authenticity is constructed – the design, placement and location are all contextually important and contribute to how the audience perceives a statue (Wang, 1999, p. 351).

Paine's memorialisation demonstrates the dialogue between local and national memory. Local memory refers to something constructed for smaller communities united by location or by ideology. Whereas national memory is fabricated to accommodate official needs and be palatable for large segments of society. In the nineteenth century, distinct factions used Paine as a symbol – a symbol of the Revolution and a symbol of radicalism – to inspire reform and give clout to radical movements. The existence of local memory is, however, exclusionary. It includes only those who have membership in the location or organisation. National memory is also exclusionary. To transcend time and remain viable in the national consciousness, historic individuals need to contribute to the unique 'American' memory. In the twentieth century, national attempts to claim Paine's memory eclipsed his radical associations. By emphasising his patriotic past through his revolutionary actions, and ignoring his radicalism, Paine became a palatable figure, even if not an entirely authentic one.

Unquestionably, memory and its perceived authenticity are constructed processes; this much is evident in Paine's legacy (Cray, 1999, pp. 575–576). Paine's historical neglect is likely due to many reasons, the most enduring of which was the publication of *Age of Reason* in 1794, amidst the religious enthusiasm that swept America during the First and Second Great Awakening (Foner, 1993, p. 203). As an attack on the authority of scripture that suggested perceived Biblical truths were superstitious illusions, *Age of Reason* was highly contentious (Andrews, 2000, p. 192). Paine's unorthodox religious position meant the vast majority of the American public classified his ideals as radical and therefore dangerous to morality and virtue (Den Hartog, 2013, p. 144). Paine's unquestionable radicalism underlined the existent controversy over his principles. In the Revolution's aftermath, from the start of Jefferson's presidency, Democrats disassociated themselves, and the Revolution, from its radical factions. The new government viewed radicalism as foreign threats to society, and as such, radical individuals and ideas found no place within the new administration. Mainstream politicians ostracised Paine and his radical ideas. Subsequent commemoration attempts reflected this exclusion (Coltar, 2007, pp. 211–213). The public did not completely forget the memory of Paine, and his actions throughout the Revolution, although the American public marginalised and sanitised his memory. Indeed, social units ensured multiple grassroots revivals of his memory by reformulating it to suit their needs. With this reformulation, an 'authentic' memory of Paine began to develop as one inextricably tied to radical segments of society.

With his death and the Revolution in the memorable past, the period from 1820–1850 was replete with conflicting memories of Paine and his works. As societal divisions grew throughout the century, historical memory and commemoration attempts became politicised. Individual groups claimed revolutionary symbols as their own, which lead to sharp divisions in the legacy of many figures (Bodnar, 1992, pp. 22, 28). As Sophia Rosenfeld explained, by the mid-nineteenth century Paine had become 'a patron saint of radicals and revolutionaries everywhere' (Rosenfeld, 2011, p. 255). Members of these radical factions that appropriated Paine were commonly active in multiple societies due to overlapping values

(Newman, 1997, p. 179). The economic conflict of the early 1820s coalesced in severe class and political tensions, as well as within a cultural forum of competition over national memory. Reformers redefined revolutionary figures as defenders of individual rights and liberties, rather than nation-builders (Bodnar, 1992, p. 27).

As early labour movements had few heroes of their own, labour reformers revived political figures like Paine, previously forgotten by the US public. Even if the appropriation of such a controversial figure made some of the leaders wary, Paine's ideology shaped and stimulated labour movements. Likewise, Paine had a strong appeal to American immigrants. As an immigrant himself, Paine, wrote about America with great promise. His deep-rooted faith in the potential of the new nation appealed greatly to those searching for an escape from European institutions (Durey, 1987, p. 196). Celebrated for his advocacy of natural rights, and a fervent belief that republicanism would safeguard equal economic opportunity to all classes, Paine and his *Rights of Man* became working-class icons (Foner, 1976, p. 264).

Regardless of the vibrant appropriation of Paine by labour organisations, much of the US public considered Paine an infidel – a moniker used to characterise those with unorthodox religious beliefs (Brown and Stein, 1978, p. 31). This degrading classification caused freethinkers to become the most vocal advocates of Paine. Like labour organisations, promotions of freethought gained traction in the early 1820s. Though many in the labour movement were not freethinkers, a significant number of freethinkers believed fiercely in societal change through labour reform and actively participated in the two communities (Voss, 1987, p. 142).

Class conflict in the 1830s also penetrated religious organisations. Evangelical recruitment during the Second Great Awakening became aggressive as they took up revolutionary rhetoric and symbols to reach more people. Participation in politics and wider cultural trends allowed preachers to 'promote righteousness and eliminate national sins' from within existing systems. Minister Peter Cartwright adapted Paine's words from the *American Crisis* when he proclaimed 'these are the times that tried men's souls and bodies too' (Andrews, 2000, p. 230). To reach the widest possible audience, Evangelicals embraced more radical institutions and their heroes. Though it was Paine's revolutionary aspects they heralded, separation of this legacy from his more controversial aspects proved impossible.

Although Paine and Evangelicalism held contradictory ideologies in many regards, Evangelicals constituted a large segment of Abolitionist communities. These communities whole-heartedly claimed Paine in their efforts to eliminate slavery. Historically, sources have credited Paine as one of the founding members of the first American anti-slavery society (Kaye, 2000, p. 51). William Lloyd Garrison's *The Liberator* frequently and enthusiastically referenced Paine's antislavery nature. Indeed, the motto on all publications of *The Liberator* between 1831 and 1836 – 'Our Country is the World – Our Countrymen are Mankind' – was Garrison's adaptation of Paine's famous saying (Sinah, 2016, p. 339). This motto called for an internationalist form of abolition to rid America of slavery.

The revival of Thomas Paine's legacy in the early nineteenth century took form in celebrations of his birthday, linked to early labour movements within the

United States (Foner, 1976, p. 264). Groups frequently held these celebrations for Paine in venues associated with working-class radicalism such as Tammany Hall, New York, where the first ceremony took place in 1825 (Claeys, 1989, p. 403). In colonial America, celebrations of monarchs' birthdays dominated festive culture. After the Revolution, presidential birthdays became the focus of celebrations. (Newman, pp. 19, 46; Maier, 1973, p. 27). Elites across the political spectrum came to dominate these early appropriations of revolutionary heritage. Consequently, in this debate over ownership of the Revolution, the working-class sought to find their own celebrations. The contest over festive culture acted as a stage for the debate over a national memory (Cray, 1999, p. 567). Birthday celebrations for Paine challenged standard public festivities by focussing on a radical figure. During the next two decades, Paine's birthday celebrations gained popularity, occurring annually, with rising attendance (Foner, 1976, p. 264).

Gilbert Vale, a prominent member of the freethought community in New York, created the impetus behind the first statue in July 1837 after he learned about the lack of commemoration for Paine at the then-annual Paine Ball. Despite economic depression and rising unemployment rates, he raised over one thousand dollars for the statue. Collectively the upper classes viewed Paine as a 'rabble-rousing atheist' and were unwilling to help fund his monument (Voss, 1987, pp. 142–143). Funds for the statue came primarily from workingmen of the city, which demonstrated the commitment of labourers to Paine (*The Beacon*, 15 July 1837). The inscription on the main face of the statue – 'Erected by Public Contribution' – noted their immense support. As demonstrated in the creation of the monument, while labour movements created the widespread admiration of Paine and paid for the monument, it owed much of the organisation and publicity to freethinkers (Foner, 1976, p. 268).

John Frazee, an admirer of Paine, sculpted the monument *pro bono* (Voss, 1986, pp. 31, 46–47). Unveiled on November 1839, Vale had the statue positioned at Paine's New Rochelle gravesite (Lause, 2008, p. 407). This location is crucial to understanding the reception of the monument. As a New York City suburb, New Rochelle was not a high-traffic area. With few exceptions, those who encountered the monument were searching for it, and more than likely were admirers of Paine. While the location is not outwardly restrictive, and the monument is not obviously hidden, its location suggests Vale built it for a particular audience. Thus, the monument is inherently reflective of the state of Paine's memory: while not hidden from the American public, only a minority viewed the monument. That minority sustained the memory of Paine for the rest of the century.

A simple design, the monument is a marble pedestal, with a decorative cap, an image of Paine carved on the front and surrounded by a laurel wreath (Figure 12.1).[2] The inscription above the image of Paine reads:

Thomas Paine
Author of
Common Sense

Figure 12.1 This is the Paine statue in New Rochelle as it appears in 2011. In 1839 when the monument was originally unveiled, it would not have had the bust on the top, or inscriptions on the sides.

Source: Copyright: Peter Radunzel, 'Thomas Paine Memorial', (2011).

The focus on *Common Sense* is notable. For the public to accept the monument, it needed to emphasise the sole universally agreeable aspect of Paine: his involvement in the Revolution. The top of the statue reads:

The World is My Country
To Do Good My Religion.
Paine's Motto

Such an inscription adds complexity to the memory disseminated. The fact that Vale did not feel it was appropriate to attribute the quote to *Age of Reason*, instead, calling it simply, 'Paine's motto', demonstrates again the controversy over Paine.

Heralded as a hero of the Revolution, yet despised by many Americans, individuals kept his memory alive, primarily labourers. The monument itself reflects such intricacies of memory.

Almost forty years later, in an attempt to participate in mainstream celebrations for the centennial of the American Revolution, freethinkers attempted to place a bust of Paine in Philadelphia's Memorial Hall Centennial exhibition, as a gift to the city. Amidst society's renewed interest in historical remembrance, they thought the monument was a fitting tribute. However, the city deemed it too controversial for display (Fruchtman, 1994, p. 441). The *New York Times* speculated that 'such was the prejudice against Paine that none would accept it' ('Place for Paine's Bust', 12 September 1905). Almost thirty years later, there was another attempt to place the bust on public display in Philadelphia, and this time the city consented. Still, the city modified the original inscriptions. The first inscription portrayed an accurate memory of Paine, but one that only a minority accepted. Although this version of Paine was authentic, the full-bodied memory of Paine was not an acceptable message for public commemoration. The two inscriptions reveal the political processes imbued in memory, as the city deemed alterations necessary to increase patriotic fervour and downplay controversy.

As seen in the processes surrounding the plaque in Philadelphia, the controversy over Paine's religion eclipsed the legacy of his contributions to the Revolution amongst the broader public (Jacoby, 2007, p. 78). Amidst an age ruled by lords of industry, fringe collectives of all kinds challenged the development of class-tension ridden Gilded Age America (Kaye, 2005, pp. 161–163). After consistent vandalism throughout the century, Vale's 1839 monument underwent repair and rededication in 1881 (Figure 12.1). Its structure received few alterations, although benefactors added multiple inscriptions ('Thomas Paine's Monument', *The Kansas City Evening Star*, 16 June 1881). Undeniably, the monument evolved alongside the memory of Paine. The changes in the inscriptions are important to note as they demonstrate a clear shift in the message the statue propagated from the original. The three previously unmarked faces displayed inscriptions from *Age of Reason* and *The Crisis*. The opening paragraphs of *Crisis I* and *Crisis X* are inscribed on the right face. The left and back face have quotes from *Age of Reason*. There are more quotes from *Age of Reason* on the statue than any of Paine's other works, which reflects the dispositions of the freethinkers who rededicated the statue. However, by adding quotes from *Age of Reason* and *The Crisis* on the other statue facades, its designers claimed versions of Paine many viewed as opposed. Paine the revolutionary hero, according to many, was separate from Paine the infidel. In bringing these two legacies together, the freethinkers who erected the statue challenged the notion that Paine could not be both a patriot and a freethinker.

In 1884, on the 147th anniversary of the birth of Thomas Paine, a group of men met at the Manhattan Liberal Club to form the Thomas Paine National Historical Association (TPNHA). The society and its radical members aimed to 'perpetuate the memory and works of Thomas Paine . . . to refute the various slanders' (Benton, 2005). The society's beliefs ran parallel to many other liberal organisations that had claimed Paine's memory in the decades previous. As the primary custodians of Paine's memory, the TPNHA created the impetus behind many subsequent forms of memorialisation to Paine in the United States. In many ways, the TPNHA became the official and authentic voice on Paine. Their efforts throughout the following decades revolved around changing the national perception of Paine, shifting his legacy from one tied only to Paine's radicalism to one that was more all encompassing. Through the Association's effort, this period also saw the first publication of an academic biography of Paine in 1892 and a collection of Paine's works in 1894. Moncure Daniel Conway, a former minister, Virginian abolitionist and TPNHA president, was responsible for these works (Burchell, 2007, p. 184; Jacoby, 2007, p. 189). Such publications were crucial in the dissemination of the memory of Paine to a broader audience, rather than just individuals involved in radical movements.

In July 1898, *The Kansas City Star* reported that a bronze bust of Paine was planned for Washington D.C. on a plot of land purchased by the TPNHA. Wilson Macdonald, one of the society's founding members, was to sculpt Paine's bust and the association, Brooklyn Philosophical Society and Manhattan Liberal Club would pay for it ('Colossal Bust of Thomas Paine', 7 July 1898). This never happened. Instead, in 1899, Macdonald sculpted a bronze bust of Paine and placed it on top of the New Rochelle monument (Figure 12.1). Whether this was a misreport from the Kansas paper, or a failed attempt to place a monument in the capital, is unclear. Undoubtedly, a national acceptance of Paine was challenging, due in part to the promotion of his religious unorthodoxy and the political leanings of those who supported the monument.

Shortly after the bust was placed on the statue, in 1905, the city of New Rochelle received ownership of the statue and rededicated it for a second time, which garnered a large audience. This event promoted the idea that the Revolution's success 'was as much due to his [Paine's] pen as to the general-ship of George Washington' ('Colossal Bust of Thomas Paine'). Speeches at the event largely ignored Paine's religious views, contrary to their prominent place on the monument itself. The celebration indicated the potency of revolutionary memory and the willingness for communities to claim controversial figures to partake in that heritage. It is clear through the day's pomp and circumstance and speech subject matters that the rededication of the monument was more about New Rochelle's connection to the Revolution via Paine, rather than Paine himself (*New York Times*, 8 May 1910).

Five years later, on Decoration Day in 1910, New Rochelle converted the Thomas Paine House into a museum. The federal government granted the building to Paine in gratitude for his wartime service (*New York Times*, 8 May 1910). Dedicating the museum on what is now Memorial Day emphasised the revolutionary heritage of Paine, and alluded to his position as a patriotic figure. The emergence of a tempered memory is evident in the rededication of the monument in 1905 and the creation of the museum in 1910. The twentieth century saw a popular revitalisation of the notion of Paine as a patriot, outside of the radical organisations that had always promoted that idea. Although many of his supporters were still radicals, the publically accepted version of Paine was that of the Revolution's propagandist.

Fittingly, the period of the Great Depression, World War Two and McCarthyism saw the most vibrant recognition of Paine. When faced with multiple national crises, the United States identified and reclaimed Paine 'as the original and persistent voice of the American spirit' (Kaye, 2005, p. 195). During the Revolutionary War, Paine's writings provided comfort to patriots weary of the war. However, this recognition was not all encompassing. Appropriations during this period exclusively focused on Paine's Revolutionary Era works and used his words for their original intention: to spark morale and nationalism. Amidst the worrying rise of European dictatorships, Americans clung more closely to domestic freedoms (Thompson, 1991, p. 86). As a fervent promoter of American freedom and liberty, Paine provided rhetorical security for Americans. The country identified and reclaimed Paine 'as the original and persistent voice of the American spirit', which made his legacy nationally acceptable for the first time and not relegated solely to the local cultures of centuries before (Kaye, 2005, p. 195).

Amid the Great Depression, President Franklin D. Roosevelt's informative radio fireside chats to the American public included a reference to Paine (Ryfe, 1999, p. 81). In one such chat in 1942, Roosevelt referenced a revolutionary heritage when he compared soldiers in 1776 and those in 1942. To encourage the nation recently drawn into war, Roosevelt reclaimed the words that Washington ordered read 'to every regiment in the Continental Army' (Franklin Roosevelt, 23 February 1942). Such rhetoric acted as an assurance to the nation. Roosevelt took the words from the first paragraph of *The Crisis I*, written by Paine in 1776:

> The summer soldier and the sunshine patriot will, in this crisis, shrink from the service of their country; but he that stands it now deserves the love and thanks of man and woman. Tyranny, like hell, is not easily conquered; yet we have this consolation with us, that the harder the sacrifice, the more glorious the triumph.
>
> (Franklin Roosevelt, 23 February 1942)

With these words, Roosevelt became the first president since Thomas Jefferson to use Thomas Paine's words in a public forum (Kaye, 2005, p. 195). In making them immediately relevant, and explicitly crediting Paine, Roosevelt shifted Paine's legacy from one previously observed by radical factions exclusively, to one that was, for the first time, nationally acceptable. When Roosevelt stated, 'So spoke Americans in the year 1776. So speak Americans today', his appropriation of *The Crisis Papers* allowed for the emergence of a 'new' hero, while simultaneously underlining a revolutionary heritage (Franklin Roosevelt, 23 February 1942). In doing so, Roosevelt not only enshrined Paine as a revolutionary hero, but also elevated Paine's words. By emphasising Paine, Roosevelt focused on soldiers and the principles for which they fought. Although the memory of Paine was still plagued with controversy, World War Two transformed him into a figure whose words held not only a distinct revolutionary heritage, but also ones

Americans could revere contemporaneously. Driven by presidential appropriation, the public accepted the notion that the authentic Paine was a Revolutionary War hero in this period.

While presidential appropriations of Paine were significant, the TPNHA facilitated all successful commemoration during the twentieth century. The Association, staying true to their original purpose, promoted and funded the initiative to restore Paine's citizenship in 1946 and two subsequent statues (Ayers, 1988, p. 185). However, without the support of more conventional sectors of society, it is possible that these efforts would never have come to fruition. Here, the dialogue between national and local memory is clear. In the case of Paine, construction of an acceptable national memory meant fixating on only one aspect of his legacy and disregarding the rest, as Roosevelt did. Roosevelt's words in 1942 resonated strongly with American soldiers, as Paine's words did during the Revolution. By 1943, military references to Paine had become routine. A US Air Force B-17F Flying Fortress named *Tom Paine* carried the inscription: 'Tyranny, like hell, is not easily conquered' echoing Roosevelt's speech (Kaye, p. 221). Frank Dobie, a Cambridge-based American history professor met the crew and commended how: 'All its crew . . . believed in Tom Paine' (Dobie, 1946, p. 46). A champion of democratic liberty and an advocate of freedom, Americans held Paine's writings, particularly *The Crisis Papers*, in utmost esteem. Such uses of Paine exhibited a negotiation of national identity from 'some remote past' in an attempt to meet the needs of the present, of a nation at war (Thelen, 1989, p. 1119).

Nevertheless, national appropriation of Paine did little to endear his legacy to more conservative factions. On 12 June 1942, the *New York Times* reported that the Fairmount Park Commission in Pennsylvania refused to erect a statue to Paine. Although the Commission conceded Paine was an 'author-hero' of the Revolution, that did not outweigh the perceived negatives of his reputation (Kaye, 2005, p. 221). Justifying the refusal, they claimed his 'reputed religious views, might make such a statue objectionable to many' ('Bar Statue of Tom Paine', *New York Times*, 12 June 1942). The years Paine spent abroad, enhanced many of concerns over Paine. In the minds of many Americans, Paine was synonymous with foreign, un-American philosophies (Jacoby, 2008, p. 42).

New Rochelle was the only location a statue of Thomas Paine existed in the United States for over a century. Given the strong ties he had to this location during his life, this is unsurprising. However, with New Rochelle as the focal point for Paine remembrance, the two statues erected outside of that location are significant. On 4 July 1950, Morristown, New Jersey unveiled the first statue of Thomas Paine erected somewhere other than New Rochelle ('Paine statue unveiled', *New York Times*, 5 July 1950).

The monument stands in Burnham Park, near the site of a number of the Continental Army's encampments during the harsh winter of 1780 (Figure 2). Before revealed, President Truman published a letter supporting the creation and dedication of the monument ('Truman aides Paine group', *New York Times*, 30 May 1950). Crucially, this was the first statue, and indeed the first period, in which the presidency publicly supported an initiative to memorialise Paine. In doing so,

Figure 12.2 The statue of Thomas Paine in Burnham Park
Source: Copyright: Dan Beards, 'Thomas Paine in Burnham Park, Morristown N.J', (2008).

Truman, like Roosevelt before him, allowed the memory of Paine to take hold in the national consciousness, for a brief time.

Unlike the previous statue to Paine, the message propagated in this was one-dimensional. Sculpted completely in bronze, the fourteen-foot statue depicts Paine writing an instalment of *The Crisis* on a drumhead, with a rifle perched atop his knee ('A heroic figure near completion', *New York Times*, 20 May 1950). The purposeful design of the monument instructs visitors on how they should understand the subject on display (Young, 1993, p. 43). Here, Paine is clearly associated with his journalistic military services during the American Revolution. The inscriptions reference no other aspect of Paine's life or writings, and quotes from his writings only exist on one face of the statue. While the figure highlights Paine, the inscriptions on the pedestal of the statue only reference Paine through other figures. In what appears to be an attempt to justify the monument and his importance, one side of the pedestal features quotes about Paine from other revolutionary figures. Indeed, though the front face of the base lists Paine's four key texts, a George Washington quote takes up significant space. While all the quotes are about the services Paine rendered the country during the Revolution, it is a very clear attempt to place him within a revolutionary context, on par with, and revered by, his already glorified contemporaries.

The statue credits no particular publication, but has five quotations inscribed; three are from *Crisis Papers*, one from *Rights of Man* and one from *Age of*

Reason, although the sentiments reflected in all of them would appear to be from his revolutionary works. It is therefore apt that this statue, devoid of other indications of Paine's works, was the first statue to him in a public area. The statue enforces a clear perception of Paine as a patriot, in a location tied exclusively to his service during the Revolutionary War and no other aspect of his life. It is an uncomplicated message, easily accessible to the American public, without a hint of controversy.

The Morristown statue was the first built somewhere other than New Rochelle. Significantly, this period is unique because groups that did not necessarily overlap, unlike the century before, appropriated Paine for the first time. The 1930s saw continued support for Paine from labour organisations, and growing connections between supporters of Paine and the Communist Party. However, by the 1940s soldiers became one of the largest forces behind commemoration, due to the image of him perpetuated by presidential speeches. National commemoration of Paine also continued, in varied form, throughout 1952. It is no coincidence that the end of this surge in national remembrance corresponded with the height of McCarthyism. After this, the memory of Thomas Paine shifted drastically.

By 1955, the positive momentum seen earlier in the decade all but disappeared. Mayor Walter H. Reynolds, of Providence, Rhode Island, refused to approve a statue, 'because Paine was and remains so controversial a character' ('Thomas Paine is Still too Controversial', *New York Times*, 23 September 1955). This rejection enflamed the American Civil Liberties Union. They argued Paine's primary focus during the Revolution was in sustaining the conflict for freedom, which they defined as: 'a struggle to establish the right of all men to speak their minds freely, without penalty or repression of government' (Amper, *New York Times*, 24 September 1955). Amidst a hostile atmosphere, individual liberties such as free speech, when exercised, aroused suspicion. This praise of Paine doubled as a thinly veiled reference to McCarthyism. It is appropriate that the memory of Paine broadcasted in the 1950s was Paine, champion of American freedom. Irrespective of the seemingly positive momentum his memory had gained throughout the century, early associations with radical factions, particularly American Communism and his controversial religious views, hindered a wider acceptance of Paine nationally during an era dominated by fears of internal subversion.

The memory of Paine propagated throughout the period focused exclusively on his revolutionary writings to make him more palatable to the American public. Much like the decades before, soldiers commonly appropriated Paine. Veterans of the Vietnam War facilitated the Winter Soldier Investigation, calling the movement's men 'winter soldiers' in a play on Paine's words from the first *Crisis Paper* (Paine, 1776). In aligning themselves to a heritage of faithful revolutionary soldiers, the movement emphasised a patriotism not previously held by war protestors. The adoption of Paine by protestors, even if they were veterans, fits with the radical claim to his legacy in the preceding century.

Diversity in the appropriation of Paine exhibited the shift in his legacy throughout the mid to late twentieth century. Once commemorated only by radical groups, his memory in this period became part of the broader national memory of the

Revolution. The 1990s were a period of cultural conflict for the United States (Gitlin, p. 1–3). Consequently, the United States saw a boom in monument construction, from 1980 to 2000, in an attempt to bind disparate factions under one common history. Breaking from previously held conservative tradition, President Ronald Reagan claimed Paine in his 1980 acceptance speech at the Republican National Convention. He attested to the exceptional character of Americans by saying Paine had them in mind when he wrote, 'during the darkest days of the American Revolution – "we have it in our power to begin the world over again"' (Ronald Reagan, 17 July 1980). This vague yet powerful and utopian appropriation of Paine was common in the 1990s (Gitlin, 1995, p. 78). Paine's writings between 1776 and 1794 all reflect a fervent belief not only in the idea of a United States, but that the new nation had the potential to be the greatest and most influential in the world. This affirming rhetoric appealed to an American nation faced with a crisis, and because of this, the version of Paine created in this period is a version of Paine that can endure.

Amidst this increase in memorialisation and frequent presidential appropriation of Paine, in 1991 there was an effort to erect a statue to Paine in Washington D.C. The *New York Times* informed its readers by a letter from Congresswoman Nita Lowey that 188 members of the United States House of Representatives co-sponsored her bill to place a statue of Thomas Paine on federal property. She boasted that the sponsorship and passage of the bill were entirely bipartisan – a joint effort with Idaho Republican Senator Steven Symms (Ames, *New York Times*, 1 May 1994). Subsequent support for the bill covered 'the political spectrum' and constituted a 'tribute to the contributions of Thomas Paine'. Editorial letters in *The New York Times* seemed generally amenable to the statue. When the bill became public law, it stated 'the TPNHA . . . is authorized to construct in the District of Columbia . . . an appropriate monument to honour the United States Patriot' (Public law 102–407, 1992). Following congressional approval of the monument, as per the Commemorative Works Act, the National Memorial Commission decided upon the statue's location (Robyn, 2016, p. 242). They held a hearing in which the TPNHA presented five scholars. An account of the hearing by historian Eric Foner revealed that the primary concern of the board was Paine's religious views. One of the first questions asked was 'Was Thomas Paine an atheist?' (Foner, 2007, p. 38). Despite the monument's unanimous approval, the overwhelming concern with Paine's religious views is noteworthy. Deemed to be of sufficient importance, Congress ultimately granted the honour of zone one status, where it would sit among the likes of Abraham Lincoln and Thomas Jefferson (Ames, *New York Times*, 1994). Approval for a Thomas Paine statue in Washington D.C., the epicentre of American historical memory, demonstrated the widespread endorsement and gratitude of Paine that grew within American politics during the twentieth century.

The terms of the aforementioned law are as significant as its passage. The bill stipulated that 'The United States shall not pay any expense of the establishment of the memorial' (Public Law 102–407). Though Congress deemed Paine important enough to merit a statue, the commemoration of him was not so crucial that the federal government would pay for, or even contribute to, its creation. Such terms were common – the Vietnam Veterans Memorial on the National Mall faced the same complication. Despite more than a decade attempting to raise funds for the Paine memorial, it ultimately failed (Straus, 2015; Robyn, p. 242). Attributing the lack of public funding solely to the controversial nature of Paine's memory would be misleading, as there are no printed records of outcry against the monument. However, the monument was not a priority. Throughout fundraising for the Paine monument, momentum was also gaining for the WWII monument. While the latter received some federal funds, public initiative raised much of the money. Though the memory of Paine had entered national consciousness, the lack of government funds demonstrated that the significance and reverence for Paine still required local cultures to preserve it.

On 7 June 1997, during a celebration called 'Occupation of Bordentown' designed to promote the town's revolutionary heritage, Bordentown, New Jersey, unveiled the most recent memorial to Paine (Figure 12.3). With strong support from the Bordentown Historical Society, a committee organised the monument's creation (Silver, 2007, p. 141). Overall, Bordentown is a place deeply proud of its history and promotes connections with figures from Napoleon to Clara Barton on commemorative plaques throughout the town. They included Paine in this remembrance as he bought property in Bordentown before leaving for Britain – the only record of Paine buying property anywhere in the United States

Figure 12.3 Paine with writings, with a gun at his feet, highlighting service to and in the Continental Army during the Revolution as a writer and Aide de Camp.

Source: Copyright: Lee B. Spitzer, 'Bordentown Thomas Paine Statue', (2017).

(Silver, 2007, p. 143). Connecting modern-day Bordentown to its historical revolutionary past became an important issue for the residents in the 1990s, in which Thomas Paine played a large part.

Lawrence Holofcener sculpted the statue (Silver, 2007, p. 143). It depicts Paine, perched on slate, which has a quote from *Common Sense* etched into the side. A rifle and his famous publications lie at his feet, and he appears to be reading *Common Sense*. Inscriptions adorn all four sides of the statue base. The front face calls Paine the 'Father of the American Revolution' contextualising Paine as inherently patriotic. An explanation of Paine's importance to the United States and his ties to Bordentown cover the other three sides. This description of Paine conveys a more wholesome picture of his time in America, while still being relatively vague. Similarly, the memory it perpetuates is not necessarily one tied explicitly to the Revolution, as was the case throughout the rest of the century. One face of the statue has a long list of Paine's accomplishments:

> Paine's words and deeds put the concepts of independence, equality, democracy, abolition of slavery, representative government and a constitution with a bill of rights on the American agenda.

This is the only quote concerning the broader contribution Paine made to the United States at all. Other faces of the statue focus on Paine's place within Bordentown, rather than the other way around:

> Paine considered Bordentown his home; it is here he invented his bridge.

Additionally, Paine's writings, which adorn other statues, are noticeably absent. The only quotation from Paine on the statue reads:

> I had rather see my horse Button in his own stable, or eating the grass of Bordentown, than see all the pomp and show of Europe – Letter from Europe 1789.

Bordentown is the focus of the quote, and it does not indicate the controversy over the man who said it, continuing the theme of the monument. The purpose of the monument is to propagate Paine's relationship to the town, not necessarily to Paine himself.

Fundraising efforts demonstrate the impressive grassroots support the monument received. Bordentown raised all the funds for the statue with little outside support. It took two years of many community balls and bake sales to raise the necessary funds (Silver, 2007, p. 142). The statue is the only monument to date that can boast such support from a community based on a common location, rather than ideology. Though conceived by a member of the TPNHA, without the efforts of the town the statue would not have been possible. In the process of raising funds for the monument, there appears to have been no controversy at all, which is telling. His legacy of unorthodoxy and the frequent appropriation of him by radical factions plagued all hitherto memorialisation attempts. This monument is evidence of a broader acceptance of Paine. While the national statue earlier in the decade failed, and the Department of the Interior hearings proved that his religious views still caused concern, the Bordentown statue demonstrated a new type of local acceptance of Paine. Previously, radical confederacies, groups on the periphery of conventional America, held his memory. The community in Bordentown hardly fits that description, instead demonstrating a small-scale, but important, acceptance of Paine by non-radical collectives within the United States.

Within the United States, many figures of the American Revolution were instantaneously solidified in historical memory, but some, like Paine, did not immediately take hold in the national consciousness. Controversy has always governed Paine's memory (Lause, 2008, p. 407). His legacy following the Revolution, as an infidel and a radical, coloured all subsequent national remembrance of him. Succeeding generations of radicals and reformers maintained Paine's memory, each utilising it in a unique way (Kaye, 2005, p. 258). The monuments and bust to Paine that exist in the United States each propagate a different version of him and the society surrounding their creation had a profound influence. Each statue to Paine is unique in design and message and portrays the version of Paine each period understood to be authentic. The statues demonstrate how his memory developed in the two centuries since his death, and how public understanding of Paine and his place within American memory evolved. The ebb and flow of regard for Paine since his death demonstrates that memories do not regulate the present; the present dictates what memories are recalled to suit its purposes (Kammen, 1993, p. 13). With each new generational crisis, Paine emerged as an enduring source of confidence and comfort. Through presidential appropriation of Paine, the possibilities of national acceptance opened. However, to be suitable for a national audience, groups reformulated the memory of Paine to include only the aspects that connected him to his patriotic past during the Revolution. Those who evoked Paine were themselves outcasts in society or small parts of larger wholes. Claiming Paine through memorialisation allowed them to claim the Revolution more broadly, and to participate in the wider revolutionary rhetoric. Although many still deem Paine controversial, the public no longer contests Paine's significance. His words, as President Obama claimed, have become 'timeless', and their meanings have touched a nerve, positively or negatively, within Americans in every period they reappeared (Baylin, 1992, p. 67; Obama, 20 January 2009).

Notes

1 Although this article only focuses on statues to Thomas Paine, it is important to note that there are places in the United States that memorialise Paine in different forms, such as the Paine Plaza in Philadelphia.
2 A 'decorative cap' refers to the carved top to the statue, where the bust sits atop. In 1839, this cap would have been the top of the statue, as those repairing the statue added the bust at a later date. // The reasoning behind the laurel wreath is unclear. Paine was not a classicist, frequently arguing against sentimental considerations of tradition in government, advocating instead for the primacy of the living (Aldridge, 1968, pp. 370–380).

Bibliography

Secondary

Aldridge, A. O. (1968) 'Thomas Paine and the Classics', *Eighteenth-Century Studies*, 1(4), pp. 370–380.
Andrews, D. (2000) *The Methodists and Revolutionary America: The Shaping of An Evangelical Culture*. Princeton: Princeton University Press.
Ashplant, T., Dawson, G. and Roper, M. (2004) 'The Politics of War Memory and Commemoration: Contexts Structures and Dynamics', in *The Politics of Memory: Commemorating War*. London: Transaction Publishers, p. 3.
Ayers, A. J. (1988) *Thomas Paine*. London: Secker & Warburg.
Baylin, B. (1992) *Faces of Revolution: Personalities and Themes in the Struggle for American Independence*. New York: Virginia Books.
Benton, G. (2014) 'The Thomas Paine National Historical Association: Freethought, Anarchism, and the Struggle for Free Speech, Part 1', *The Truth Seeker*, September December, pp. 17–21.
Bodnar, J. (1992) *Remaking America: Public Memory, Commemoration, and Patriotism in the Twentieth Century*. Princeton: Princeton University Press.
Brown, M. and Stein, G. (1978) *Freethought in the United States of America*. Westport, CT: Greenwood Press.
Burchell, K. (2007) 'Birthday Party Politics: The Thomas Paine Birthday Celebrations and the Origins of American Democratic Reform', in King, R. F. and Begler, E. (eds.) *Common Sense for the Modern Era*. San Diego: San Diego State University Press, pp. 174–190.
Claeys, G. (1989) *Thomas Paine: Social and Political Thought*. Boston: Unwin Hyman Inc.
Coltar, S. (2007) *Tom Paine's America*. Charlottesville: University of Virginia Press.
Coltar, S. (2013) 'Conclusion: Thomas Paine in the Atlantic Historical Imagination', in Newman, S. and Onuf, P. (eds.) *Paine and Jefferson in the Age of Revolutions*. Charlottesville: University of Virginia, pp. 277–296.
Cray, R. (1999) 'Commemorating the Prison Ship Dead: Revolutionary Memory and the Politics of Sepulchre in the Early Republic 1776–1808', *William and Mary Quarterly*, 56(3), pp. 565–590.
'Den Hartog, J. (2013) 'Trans-Atlantic Anti-Jacobinism: Reaction and Religion', *Early American Studies*, 11(1), pp. 133–145.
Durey, M. (1987) 'Thomas Paine's Apostles: Radical Emigres and the Triumph of Jeffersonian Republicanism', *The William and Mary Quarterly*, 44(4), pp. 661–688.
Foner, E. (1976) *Tom Paine and Revolutionary America*. New York: Oxford University Press.
Foner, E. (1993) 'Tom Paine's Republic: Radical Ideology and Social Change', in Young, A. (ed.) *The American Revolution: Explorations in the History of American Radicalism*. Chicago: Northern Illinois University Press.
Foner, E. (2007) 'Thomas Paine and the American Radical Tradition', in King, R. F. and Begler, E. (eds.) *Common Sense for the Modern Era*. San Diego: San Diego State University Press King, pp. 38–59.
Fruchtman, J. (1994) *Thomas Paine Apostle of Freedom*. New York: Four Walls Eight Windows.
Gitlin, T. (1995) *The Twilight of Common Dreams: Why American Is Wracked By Culture Wars*. New York: Owl Books.

Jacoby, S. (2007) 'The Religious Radicalism of Thomas Paine: Why the Age of Reason Still Threatens Unreason', in King, R. F. and Begler, E. (eds.) *Common Sense for the Modern Era*. San Diego: San Diego State University Press, pp. 77–79.
Kammen, M. (1978) *Season of Youth: The American Revolution in Historical Imagination*. New York: Alfred Knopf.
Kaye, H. (2000) *Thomas Paine: Firebrand of the Revolution*. Oxford: Oxford University Press.
Kaye, H. (2005) *Thomas Paine and the Promise of America*. New York: Hill and Wang.
Keane, J. (1996) *Tom Paine: A Political Life*. London: Bloomsbury Publishing.
Lause, M. (2008) 'The "Unwashed Infidelity": Thomas Paine and Early New York City Labor History', *Labor History*, 27(3), pp. 385–409.
Lowen, J. (1999) *Lies Across America*. New York: New York Press.
Maier, P. (1973) *From Resistance to Revolution: Colonial Radicals and the Development of American Opposition to Britain, 1765–1776*. London: Routledge.
Marschall, S. (2013) 'Collective Memory and Cultural Difference: Official vs. Vernacular Forms of Commemorating the Past', *Journal of South African and American Studies*, 14(1), pp. 77–92.
Newman, S. (1997) *Parades and the Politics of the Street: Festival Culture in the Early Republic*. Philadelphia: University of Pennsylvania Press.
Robyn, R. (2016) 'Erasure of Public Memory: The Strange Case of Tom Paine in Washington, DC', in Cleary, S. and Stabell, I. L. (eds.) *New Directions in Thomas Paine Studies*. New York: Palgrave Macmillan, pp. 229–245.
Rosenfeld, S. (2011) *Common Sense: A Political History*. Cambridge: Harvard University Press.
Ryfe, D. (1999) 'Franklin Roosevelt and the Fireside Chats', *Journal of Communication*, pp. 80–109.
Silver, M. K. (2007) *In His Footsteps*. Bordentown: Hometown Press.
Sinha, M. (2016) *The Slave's Cause: A History of Abolition*. New Haven: Yale University Press.
Straus, J. (2015) 'Monuments and Memorials Authorized Under the Commemorative Works Act in the District of Columbia: Current Development of In-Progress and Lapsed Work', *Congressional Research Services*.
Thelen, D. (1989) 'Memory and American History', *Journal of American History*, 20(1), pp. 1117–1129.
Thompson, T. (1991) 'The Resurrection of Thomas Paine in American Popular Magazines', *Midwest Quarterly*, 33, pp. 75–92.
Voss, F. (1986) *John Frazee, Sculptor*. Washington, DC: Boston Athenaeum Library.
Voss, F. (1987) 'Honouring a Scored Hero: America's Monument to Thomas Paine', *New York History*, 68(2), pp. 133–150.
Wang, N. (1999) 'Rethinking Authenticity in Tourism Experience', *Annals of Tourism Research*, 26(2), pp. 349–370.
Young, A. (2000) *The Shoemaker and the Tea Party: Memory and the American Revolution*. Boston: Beacon Press.

Primary

Presidential speeches

Obama, B. (2009) *Inaugural Address*, 20 Jan., Peters, G. and Woolley, J. (eds.) *American Presidency Project*. Available at: www.presidency.ucsb.edu/ws/index.php?pid=44.

Reagan, R. (1980) *Address Accepting the Presidential Nomination at the Republican National Convention in Detroit*, 17 Jul., Peters, G. and Woolley, J. (eds.) *American Presidency Project*. Available at: www.presidency.ucsb.edu/ws/?pid=25970.

Roosevelt, F. D. (1942) *Fireside Chat*, 23 Feb., Peters, G. and Woolley, J. (eds.) *American Presidency Project*. Available at: www.presidency.ucsb.edu/ws/index.php?pid=16224&st=Paine&st1=.

Laws

'Public law 102–407' (1992) 102 Congress, 13 Oct. Available at: http://thomas.loc.gov/cgibin/query/D?c102:5:./temp/~c102GwqEDC.

Newspaper article

Ames, L. (1994) 'Progress on the Paine Memorial', *New York Times*, 1 May.

Amper, R. (1955) 'Rejection of Tom Paine Statue Inflames Civil Liberties Union', *New York Times*, 24 Sept.

The Beacon (1837) 15 Jul.

Kansas City Evening Star (1881) 'Thomas Paine's Monument', 16 Jun.

Kansas City Star (1898) 'Colossal Bust of Thomas Paine', 7 Jul.

New York Times (1905) 'Place for Paine's Bust', 12 Sept.

New York Times (1950) 'A Heroic Figure Near Completion', 20 May.

New York Times (1950) 'Truman Aides Paine Group', 30 May.

New York Times (1950) 'Paine Statue Unveiled', 5 Jul.

New York Times (1955) 'Thomas Paine is Still Too Controversial', 23 Sept.

New York Times (1955) 'Mayor Backs Action on Paine Memorial', 25 Sept.

New York Times (1992) 'Letter to the Editor: Thomas Paine Statue Coming to Capital', 31 Jan.

New York Times (1992) 'Letter to the Editor: No Brotherly Love Yet for Thomas Paine', 19 Feb.

Personal accounts

Dobie, F. J. (1946) *A Texan in England*. London: Hammond and Company.

13 Familiarity breeds content

Shaping the nostalgic drift in postbellum plantation life-writing

David Anderson

Introduction

First published in 1887, Susan Dabney Smedes's (1887, p. 3) *Memorials of a Southern Planter* was written so that her father's grandchildren could save for posterity "the memory and example of his life" as a slaveholding planter in Virginia and Mississippi during the antebellum period. "They will come to mature years in a time when slavery will be a thing of the past," wrote Smedes.

> They will hear much of the wickedness of slavery and of slave-owners. I wish them to learn of a good master: of one who cared for his servants affectionately ... and with a full sense of his responsibility. There were many like him.

Smedes, a doughty defender of slavery as an institution, was asserting her determination to hand down her father's name untarnished, his reputation imperishable (Dabney, 1978, pp. 2–3; *Daily Times*, 1889, p. 8). Occupying an exalted place in the plantation calendar, Christmas on the Dabney's plantation made the point well. Smedes appropriated Christmas and its rituals and used them to celebrate her father's deeds of benevolence, his generosity of spirit evidenced in tokens of esteem and flattery distributed to his dependents. In one festive scene we encounter Dabney concocting eggnog for "his former overseer and other plain neighbours" and singing songs to servants and field hands, an intimate portrait of plantation community life in which social hierarchies are overcome by a "symbolic gesture of deference" to lower social classes (Smedes, 1887, p. 160; Nissenbaum, 1996, p. 264). The Old South Christmas is one example of the carefully designed literary strategies used by elite whites to elicit nostalgia for an imagined plantation community (Anderson, 2014). Moreover, the vignettes of plantation life celebrated by Smedes present a re-telling of those landscapes from which slavery is left only as memory with the intention of fostering an authentic, real, natural, and thus true formulation of the Old South. With this tactic, a stable, pre-industrial pastoral of faithful slaves and harmonious race relations elides the traumatic history of slavery, shifting attention from the racial discrimination and violence of the Jim Crow era.

Discussion

Beginning in the 1880s, members of the old planter elite published autobiographies, memoirs, and other reminiscences to describe their experiences of life on the plantation in slavery times, reaching its zenith during the first decade of the twentieth century. In addition to Susan Dabney Smedes, its authors include R.Q. Mallard, Letitia M. Burwell, John S. Wise, James Battle Avirett, Virginia Clay-Clopton, H.M. Hamill, Sara Agnes Rice Pryor, Eliza Ripley, J.G. Clinkscales, and Susan Bradford Eppes. Among other publishing houses, New York firms including Macmillan, Scribner's, Appleton, and Doubleday Page printed, marketed, and distributed books of plantation reminiscences to reading audiences across America, part of a regional literature, along with history and fiction, that contributed to the literary climate for a prelapsarian South (Cox, 2011, p. 108). Along with these houses, Confederate memorial groups promoted plantation memoirs and reminiscences to their memberships, contributing to the "popular and critical success" of the genre. Such was the deluge of these volumes by the turn of the century, fears were raised among authors and publishers about oversaturating the literary market (Gardner, 2004, pp. 130–131).

Close reading of plantation autobiographies and memoirs, of the sort focused on in this chapter, reveals the ways in which a lifetime's memories are filtered and reframed, providing a mnemonic template upon which certain cultural assumptions about the Old South plantation imaginary are overlaid. Approached thus, these self-representational texts serve to open significant research questions around their content, dissemination, and reception, as well as the pervasive potency of nostalgia that such narratives possess. For a number of critics, nostalgia plays a significant shaping role in constructions of the past, an emotional and political activity expressing the needs of the individuals or social groups whose interests it serves, often the economic and political elite. Aroused by a profound dissatisfaction and uneasiness with a changing present, these memory projects reaffirm and codify invented traditions, of which Eric Hobsbawm (1983, pp. 1–14) has written, and work to establish a sense of continuity with the past, authenticating and legitimating important values, norms, and customs. More recently, a growing body of scholarship has begun to draw attention to the ways in which nostalgia for the past underpins ideas of authenticity and identity, using a cross-disciplinary approach to highlight contexts for imaginings of home, place, and landscape created by temporal and spatial dislocation and distance (Williksen and Rapport, 2010).

In his important study of southern history in the years between Reconstruction and the turn of the twentieth century, Edward Ayers (1992, p. viii) describes the New South as an "anxious place," a land whose people, looking to a future haunted by ghosts of the past, scrambled to maintain a coherent sense of themselves. Retreat into nostalgic recollections of a happier past was a common response to the New South's "unsettling dynamism," not least in the pages of plantation autobiographies and memoirs (Blight, 2001, p. 222). Bemoaning the "energetic spirit of a new age," Mrs. N.B. De Saussure (1909, pp. 9–10) throughout her memoir

Old Plantation Days repeatedly addresses "the South as it used to be," turning recollections of "happy plantation days" into tokens of "loving memory," a legacy to future generations of southern whites from one "now passing away." Like other postbellum plantation memoirists, De Saussure thus confronts the inevitable fading of things that manifest nostalgia for the Old South, presenting herself as an authentic voice of lived experience and memory, someone with "a true knowledge" of slavery days on the plantation, "whose life was spent amid those scenes." De Saussure's self-identity was rooted in the past, a legacy of history and memory.

Recent critical interventions into debates on southern autobiography and memoir draws attention to the ways in which male and female writers have used shared cultural and historical experiences to define and understand themselves. Challenging assumptions about "faulty or selective memories, conscious or subconscious agendas, or overly imaginative enhancements" that tend to frame scholarly inquiry into autobiography and memoir, John Inscoe (2011, p. 9) has explored the genre's capacity to apprehend southern history and experience by those who lived it, helping us understand the social forces that have shaped southern identity and regional distinctiveness over time. While much has been written on the canonical works of southern autobiography, Inscoe's principal focus, postbellum plantation life-writing, remains under studied in the literature, but is no less important. All too easily dismissed because of their overt mawkishness and chauvinistic positions, autobiographies and memoirs by whites during antebellum times and the racial patterns of the plantation are worthy of serious study. There exists a need to improve knowledge of this sentimental *oeuvre*, one that has yet to overcome historians' suspicion of the subjective operations of the human mind and the source material's innate tendencies toward nostalgia. In particular, more attention needs to be given to the forms and functions nostalgia takes in response to change and transformation, to demonstrate the value of restoring nostalgia to a broader context of debates concerning understandings of the self and social identities.

With this in mind, this chapter is divided into two parts. The first part begins with a brief discussion of how nostalgia creates identity continuity to offset circumstances of change and uncertainty, placing that discussion within the white South's account of its past in the aftermath of the Civil War. Drawing on a rich, but largely unexplored, body of autobiographical writing by former slaveholders and their kin, I then bring into focus the narration of the Civil War as a rift in the fabric of time, a temporal rupture that violently separated past from present. Underpinning this sense of historical discontinuity was a gnawing awareness among elite white southerners that the remembered past was fading from living memory. Autobiographies and memoirs of life on the plantation, written primarily to give such recollections a more permanent form, preserved the Old South in the aspic of nostalgia. In such a context, nostalgia is projected onto the receding past in articulation of narrative strategies and categories that forge cultural authenticity and identity, the discussion of which adds a new dimension to current debates around the Old South imaginary and its purveyors. The chapter's second section explores some of the assumptions underlying the image of the faithful slave in postbellum plantation life-writing. Here, I argue evocations of close affinities between paternal masters and loyal retainers who are content in their enslavement, set uneasily against the backdrop of Jim Crow, were part of a rhetorical strategy designed by whites to outline a future for southern race relations that adhered to the hierarchical patterns of slavery. The nostalgic conceit, which abandons the present for the past, allies with the desire to retrieve – or revive – an earlier, presumably more authentic time.

****

Building on the seminal sociological work of Fred Davis (1979), recent critics have nuanced understanding on the lineaments of nostalgic experience, evocation, and reaction, paying detailed attention to why, how, and with what effect nostalgia is employed by the individual and the group in the construction and maintenance of identity. As Nadia Atia and Jeremy Davies (2010, p. 184) argue, referencing Davis, "nostalgia serves as a negotiation between continuity and discontinuity: it insists on the bond between our present selves and a certain fragment of the past, but also on the force of our separation from what we have lost." In this context, the identity dislocations produced by abrupt social change and upheaval, and their attending anxieties and uncertainties, as Davis (1979) identified, reveal some of the ways in which the timbre of nostalgia's cognitive state acts as a compelling interpretive lens through which the past is viewed, filtered, and interpreted according to the needs of the present.

The nostalgic turn of white southerners after the Union's defeat of the Confederacy in 1865 and efforts to memorialise and commemorate the Lost Cause has attracted considerable scholarly attention over the years, much of which focuses on the complex interplay between remembering and forgetting in themes of race, reunion, and reconciliation (Janney, 2013; Cook, 2017). This corpus of literature offers valuable insight into the rhetorical processes at work during the post-war years that shaped southern identities, histories, and narratives in important ways. These processes explain, in part, the burgeoning "reminiscence industry" of the late nineteenth century to which David Blight (2001, p. 166) refers. Accounts of the antebellum era by members of the old planter class regard the Civil War as a moment of temporal rupture, an epochal shift signalled by the dual transition from peacetime to wartime and back again. Ruined by war, humiliated by defeat, and insulted by Reconstruction's radical Republican governments, southern whites turned eagerly to nostalgia for consolation, harking back to the good old days with "exaggerated tenderness" (Logan, 1950, p. 234). In his insightful history of southern identity, James Cobb (2005, pp. 73–74) rightly observes the Civil War and Reconstruction provided white southerners with a distinctive historical experience. Temporal lines were redrawn in the aftermath of the Civil War, which "fast-forwarded the antebellum southern order through the process of ageing and historical distancing," and transformed the South prior to 1860 into the Old South, "frozen away in some distant corner of time and accessible only through the imagination." Cut off from the present, the Old South became lost in time and, for that very reason, timeless – static, fixed, and immutable, "distance," in Richard Gray's (1986, p. 89) words, "could give a romantic blur to everything."

In 1926, aged 80, Susan Bradford Eppes (1926, p. 13) published *Through Some Eventful Years*, an account of her upbringing in antebellum Leon County and the Florida cotton belt, a perspective shaped in the decades that followed the events the memoir attempts to explain. Her childhood and youth on Pine Hill, her father's plantation, was "an era of 'House Parties,'" we are told, especially at Christmas time when "colored servitors, enough for every demand," readied, cooked, and served "all the luxuries of life." However, the Civil War "devastated" the region, hastening change with such "lightning-like rapidity" that white southerners "dared not look the future in the face." Old South civilisation, she concluded with sorrow, was "slowly but surely dying," emphasising a temporal dichotomy between "the traditions, the principles, [and] the customs of bygone days," an era "forever cherished," and the "dark days of Reconstruction" that followed. "Where once wealth abounded, poverty stalked, gaunt and bare" (Eppes, 1926, pp. 325, 340, 344, 370). Not even nostalgia was as good as it used to be.

The Civil War shattered the plantation world as the old order had known it. Southern whites who figured that past as a Golden Age were particularly alert and sensitive to what was lost and irrecoverable, a verity evident in their life-writing and reluctant farewell to Old South civilisation. "These were the halcyon days of the South, gone never to return," claimed Frank Montgomery (1901, p. 20), a former cotton planter, in his *Reminiscences of a Mississippian in Peace and War*. Virginian H.M. Hamill (1904, pp. 6, 38), "born in and of the Old South," voiced his "abiding regret" that the plantation world of his childhood and youth was "gone forever." In her posthumously published *Social Life in Old New Orleans*, originally a series of newspaper articles which appeared in the city's *Times-Democrat*, Eliza Ripley (1912, pp. 263, 191) grieved over "days that are gone, never to be lived again," the end of a way of "life that died and was buried fifty years ago."

The envisioned nullification of the past, of an era that was "gone," caused both distress and dislocation, and gave rise to feelings of temporal rupture and fragmentation in numerous postbellum works of life-writing. In his memoir of plantation life in McIntosh County, Georgia, Edward J. Thomas (1923, p. 5) noted that he "lived in two distinct periods of our Southern history," for the Civil War "completely severed the grand old plantation life" from the "striving conditions that followed." Visiting Burleigh, the Dabney family plantation, after the war, a family friend likened scenes there to "an evil dream," remarking unhappily, "times are changed" (Smedes, 1887, p. 241). Edward Spann Hammond (Hammond, cited in Clay-Clopton, 1905, p. 212), son of the affluent coastal South Carolina planter James Henry Hammond, felt like he "had been in two worlds, and two existences, the old and the new ... so thorough has been the upheaval and obliteration of the methods and surroundings of the past." Thus the Civil War was framed as a precipitous moment of fundamental rupture, split, and division, estranging the past from the present.

In accordance with Fred Davis's (1979, pp. 49, 102) assumption that "rude transitions rendered by history" can jar our being and mark our epochs, tropes of ruin inflect plantation memoirists' reading of the Civil War. In *My Day: Reminiscences of a Long Life*, Sara Pryor (1909, p. 273), wife of Congressman Roger A. Pryor, of Virginia, claimed she and her husband "found it almost impossible to take up our lives again" following the Civil War. "All the cords binding us to the past were severed, beyond the hope of reunion." Reunited at war's end, the Pryors "sat silently" looking out on war-ravaged Richmond, onto "a landscape marked here and there by chimneys standing sentinel over blackened heaps," where neighbours, now displaced, had once "made happy homes." North Carolinian Mary Norcott Bryan ([1912], pp. 25–26) returned home at the end of the war to find Woodlawn, the family's "beautiful and valued" home, "an abandoned plantation," the slave cabins, barns, and outhouses dismantled, trees felled, and livestock scattered. One memoirist later summed up the situation thus: "The war was our undoing, our alpha and omega" (Winston, 1937, p. 5).

The presentation of "dreadful days" of "war and fire and famine" is juxtaposed with the sentimental portrayal of life on a South Carolina lowland plantation during the 1840s and 50s in De Saussure's (1909, p. 10) *Old Plantation Days*, "the recollection of which causes my heart to throb again with youthful pleasure." Written as a letter to a granddaughter raised in the North, De Saussure recounts the Old South's "delightful open-hearted, open-handed way of living" as a rebuff to critiques of slavery, taking aim at *Uncle Tom's Cabin* (1852), Harriet Beecher Stowe's anti-slavery novel (1909, pp. 53, 17). In his memoir, *The End of an Era*, J.S. Wise (1899, p. 77) also found "outrageous exaggerations" in Stowe's novel, while James Battle Avirett (1901, p. 64) deemed it an "ignorant compend of anger, hatred and malice." Unsurprisingly, postbellum plantation romancers rejected any counter-narratives that contested their claim to an authentic past, drawing inspiration from memories of lived experience and the nostalgia of reminiscence deepened with the passage of time. Elite white southerners' stories of the southern past were more Uncle Remus than Uncle Tom.

As David Blight (2001, p. 222) has shown, romanticised accounts of the plantation and its inhabitants offered white southerners and other Americans "emotional fuel and sustenance," a counterpoint to the social and economic problems of the Gilded Age. Given Blight's emphasis on Gilded Age insecurity, it is worth pointing out that many white northerners eagerly embraced the "idyll of the Old South's plantation world" and its celebration of reunion and reconciliation in the North-South relationship. The nostalgic turn to the antebellum South during the late nineteenth century at once celebrated a claimed authenticity and hailed a region carefully attuned to the pastoral traditions of pre-industrial America; "an unheroic age could now escape to an alternative universe," one where traditional race, class, and gender hierarchies persisted (Blight, 2001, pp. 211, 222; Cox, 2011, pp. 3–5).

****

Received by critics as "faithful to fact" (*Congregationalist*, 1895, p. 527), *A Girl's Life in Virginia Before the War*, by Letitia M. Burwell (1895, pp. 12–14), traces life in the Piedmont where the author's distinguished ancestors, for nine generations, owned plantations. Bound in a light-brown cloth with a cotton bud decoration on the front cover, the volume includes sixteen black-and-white illustrations depicting Old South scenes and settings by artists William A. McCullough and

Jules Turcas. In one sketch, an elderly ex-slave, dressed in rags and carrying a walking stick, is seen talking to the mistress of the plantation, pleading to be taken in and looked after; it includes a strap line that reads, "I don't want be free no mo." The old man is directed to the plantation kitchen where he is given something to eat. Freedom after emancipation, as perceived within Burwell's reminiscences, is imagined as a "very dreadful and unfortunate condition of humanity." By way of gratitude to the plantation mistress for her "kindly" offer of food and shelter, the freedman "entertained" the Burwell family with "pleasant reminiscences" of old times, occasionally offering a "sigh that the days of glory had departed." Recounting the former slave's personal reminiscences, which "had a certain charm," Burwell observed a "mournful contrast between past and present" on the plantation and the relationships it fostered between both races, now irrevocably altered or altogether lost. As David Blight (2001, p. 286) put it succinctly, "white Southerners strove to convince themselves that emancipation had ruined an ideal in race relations."

The illustration in Burwell's sentimental volume illuminates several key themes in the faithful slave narrative, not least the assumption that slaves never really wanted to be set free. In *Dixie After the War*, Myrta Lockett Avary (1906, pp. 183–185) recalls the night her father, a Virginia planter, told his slaves that they were to be given freedom. Standing on the porch beside her father, Avary looked "out on the sea of uplifted black faces" as the momentous news was read by candlelight. "They listened silently" to their master's words, we are told, trying to make sense of a unique historical moment. "Some wiped their eyes, and my father had tears in his," wrote Avary. As the freedpeople passed before their old master, "one and all" stated their determination to stay on the plantation and carry on living and working as before. Ultimately, the impact and consequences of emancipation were far-reaching, and fundamentally reshaped the region's agricultural labour systems and working arrangements. Yet here was a smooth transition from slavery to freedom, a passionate advocacy of "white nobility, black humility, mutual obligations, faithful service, and the extended family unit – black and white" (Litwack, 1980, p. 192).

The image of the loyal slave is ubiquitous in postbellum plantation autobiographies and memoirs, as it was in turn of the century novels, plays, minstrel shows, and popular songs. Many late-nineteenth-century literary immortalisers of the Old South had, of course, experienced plantation life and witnessed slavery as children. As Grace Elizabeth Hale (1998) and Jennifer Ritterhouse (2006) have demonstrated, childhood was the crucial formative period when southern whites learned racial difference, representation, and hierarchy. "I am no apologist for slavery," wrote Eliza Ripley (1912, p. 192) in her memoir, "but we were born to it, grew up with it, lived with it, and it was our daily life." Here identity connects with place, the plantation. It is a claim that relies on the sense of authenticity conveyed by lived experience, memory, and nostalgia. Such motifs accord with a broader sentimental ideology of childhood innocence, which, as Robin Bernstein (2011, p. 4) has argued, tended to reinforce white racial projects, making them "appear natural, inevitable, and therefore justified." For romantic depicters of old plantation days, as Hale (1998, p. 54) explains, "desire for childhood innocence converged with a regional longing for racial harmony."

During slavery, white children made regular visits to the slave quarters. Their after-the-fact nostalgia is signalled through recollections of children at play and other amusements in which whites and blacks interacted. Mary Ross Banks (1882, pp. 19–20), from a white upper-class background in Georgia, recalled spending "some of the purest, happiest hours" learning "the best lessons of my life" from old Granny Sabra, whose "devotion" to the slaveholding family was "quite beautiful." Victoria V. Clayton's (1899, p. 23) *White and Black Under the Old Regime*, a collection of vignettes embracing plantation life on the banks of the Chattahoochie River, similarly delights in recounting childhood memories of visits and chats with the enslaved peoples on her father's plantation. Letitia Burwell (1895, pp. 2–3), who enjoyed a gloriously privileged childhood, remembered regular outings to the slave cabins, located about a mile from the big house, "on which occasions no young princesses could have received from admiring subjects more adulation" than she and her sister received from their "dusky admirers." That bonds of affection and mutual respect between whites and blacks ran deep, so imperfectly understood by those who had never experienced it, is stated explicitly by J.G. Clinkscales (1916, pp. 35–36) in his account of his plantation childhood which records playful scenes with Unc' Essick, the plantation foreman, a "faithful slave," a "patient teacher," and a "colored gentleman." For southern whites, such reciprocal friendships and cares, formed on plantation estates during slavery, marked a watershed in the history of southern race relations.

Raised on a rice plantation in Liberty County, Georgia, R.Q. Mallard (1892, p. vii) likened the attachments between master and slave to the "strength of the ties which connect dear kindred." Mallard was keen for his reminiscences to be read as a blueprint for the future of race relations in the South, to "contribute to the restoration of the mutual relations of kindness and confidence characterising the old regime" that had been "sorely strained" by the "unhappy" trials of emancipation, war, and Reconstruction policies. For Mallard, a return to the best traditions of the master-slave relationship as it existed within a Christian plantation community would bring about more harmonious race relations in the postbellum era and solve the South's race problem. These narrative spaces which accommodate portrayals of the master-slave dialectic, trading on assumptions of unbroken inheritances from the Old South, formulate a notion of authenticity that derives its legitimacy from some measure of imagined continuity between past and present.

Many plantation memoirists ventured similar testimony to elevate and embellish an institution that died with emancipation, animating a central tension between an authentic past – one that is anchored in its certainties – and a present that is deemed to be unmoored and off-course. Writing of his memories on Alabama's "great plantations, in their picturesque colors," H.M. Hamill (1904, pp. 31–32) insisted that the state's enslaved people were "well fed and clothed," were "moderately worked," and lived a "careless, heart-free life." The enslaved population appeared to want for nothing. Hardship and suffering came with freedom. The "care-worn faces" of the "old-time negroes" told its own story, according to

Hamill; for them, the half-century following emancipation was "full of heartache and worry," as freedom "had proven a cheat and a snare." Antebellum days served as a model against which to contrast blacks "born and trained under slavery," a group who "commands respect in the South to-day," with "those who have known nothing but freedom," an "unsatisfactory body of people generally." Black men and women born and raised in slavery on the plantation, trustworthy, dignified, restrained, "docile and reverent," would "always" be "the friend of the Southern white gentleman and lady," he claimed. Many southern whites held fast to the paternal ideal, convinced that the best elements of the old institution were the best hope for the continuance of racial order in the New South. By the turn of the century, however, this crisis of mastery had developed into a system of legalised racial segregation, voter disfranchisement, and endemic violence against blacks.

In this context, writers found it easy to retreat from worsening race relations within segregation to images that invoked a simpler, less complicated past. Lectures on plantation life and recitals of dialect stories set in a plantation setting were especially popular on the American lyceum and Chautauqua circuits during the late nineteenth and early twentieth centuries. Many of these narrations, which create tensions between authenticity and artifice, were published subsequently in literary magazines and edited collections. Martha Gielow's (1898, 1902) dialect monologues, character sketches, and "darky" songs, performed in America and Britain and published as *Mammy's Reminiscences and Other Sketches* and *Old Plantation Days*, were well received by the press. *The Atlanta Constitution (1901, p. 9)* considered Gielow's portrayals of life in the slave quarters authentic and faithful accounts; "they are life photographs of the real thing, or, better still, pictures from out your own memory and experience." Gielow, a figure almost forgotten today, began her public career as a reader of regional dialect stories written by Joel Chandler Harris ("Uncle Remus") and Thomas Nelson Page ("Marse Chan") before being persuaded to write of her own recollections growing up in Alabama's Black Belt. Interestingly, Gielow (1898, pp. vii–viii) criticised the "grossly exaggerated and caricatured" image of blacks in minstrel shows and other productions that rendered the slaves "unrecognizable by those who knew and loved them." Inspired by "actual happenings related to me by my own black Mammy," Gielow's "character sketches and jingles" presented "a correct and natural dramatic impersonation of the old-time Mammy and Daddy, the devoted foster parents to the children of the South." If the use of dialect gave plantation autobiographies and other forms of reflective writing "a perceived authenticity," then it also enhanced images of a bucolic South at a time when the Jim Crow system was expanding and racial violence increasing (Cox, 2011, p. 118).

****

Autobiographies and memoirs by former slaveholding planters and their families exhibit a conspicuous nostalgia for life on the southern plantation before the Civil War, especially the everyday personal relationships between the master class and the enslaved. Susan Dabney Smedes's *Memorials of a Southern Planter*, for example, recalls the Old South of her childhood and youth in a series of evocatively rendered vignettes that sustain and support an idealised past of paternalistic race relations that emancipation and war either ended or considerably weakened. Smedes's reconstruction of the Old South, of nostalgic reminiscences affectionately told, served as a primary vehicle through which plantation memoirists negotiated a sense of belonging in an era of transformative socio-political shifts. In short, memories were brought to bear in pursuit of an authentic regional identity – one that looked to the past for its sense of cultural distinctiveness.

Given nostalgia's major role in shaping, developing, and proliferating an authentic cultural memory of the Old South and the Lost Cause of the Confederacy, it is surprising that the topic has not received more attention from historians. Plantation autobiographies and memoirs, frequently dismissed as trite and inconsequential, are an important, though overlooked, category of Gilded Age writing projects that imagined a southern past of plantations and racial harmony. Seen, of course, through the optic of white experience, these narratives legitimated Jim Crow and justified white supremacy, ensuring blacks remained socially, politically, and economically subservient to the white ruling class. At the same time, many black writers challenged the nostalgic creed of the Lost Cause. Here, the Old South and plantation era slavery are understood not in terms of loss, regret, and reverential memory, but as a tragedy of American history, a history of oppression, suffering, and prejudice. However, for those southern whites who turned to antebellum times on the plantation, clinging to a version of the past imagined as authentic, one that sanctioned their racial privilege and power, the inclination of the nostalgic drift was to a vanished world of faithful slaves in devoted and lifelong service. Refuge was to be found in the long shadow of the Old South's sunset glow.

Bibliography

Anderson, D. (2014) 'Nostalgia for Christmas in Postbellum Plantation Reminiscences', *Southern Studies*, 21, Fall–Winter, pp. 39–73.
Atia, N. and Davies, J. (2010) 'Nostalgia and the Shapes of History: Editorial', *Memory Studies*, 3(3), pp. 181–186.
Atlanta Constitution (1901) 'Mrs. Gielow's Recital Was Splendid Success', *Atlanta Constitution*, 22 Mar., p. 9.
Avary, M. L. (1906) *Dixie After the War*. New York: Doubleday, Page.
Avirett, J. B. (1901) *The Old Plantation: How We Lived in Great House and Cabin Before the War*. New York: F. Tennyson Neely.
Ayers, E. L. (1992) *The Promise of the New South: Life After Reconstruction*. New York: Oxford University Press.
Banks, M. R. (1882) *Bright Days in the Old Plantation*. Boston: Lee and Shepard.
Bernstein, R. (2011) *Racial Innocence: Performing American Childhood from Slavery to Civil Rights*. New York: New York University Press.
Blight, D. W. (2001) *Race and Reunion: The Civil War in American Memory*. Cambridge: Belknap Press of Harvard University Press.
Bryan, M. N. ([1912]). *A Grandmother's Recollection of Dixie*. New Bern, NC: Owen G. Dunn.
Burwell, L. M. (1895) *A Girl's Life in Virginia Before the War*. New York: Frederick A. Stokes.

Clay-Clopton, V. (1905) *A Belle of the Fifties: Memoirs of Mrs. Clay of Alabama, Covering Social and Political Life in Washington and the South, 1853–66. Put into Narrative form By Ada Sterling*. New York: Doubleday, Page.

Clayton, V. V. (1899) *White and Black Under the Old Regime*. Milwaukee: Young Churchman.

Clinkscales, J. G. (1916) *On the Old Plantation: Reminiscences of His Childhood*. Spartanburg, SC: Band and White.

Cobb, J. C. (2005) *Away Down South: A History of Southern Identity*. Oxford: Oxford University Press.

Congregationalist (1895) 'Stories', *Congregationalist*, 4 Apr., p. 527.

Cook, R. (2017) *Civil War Memories: Contesting the Past in the United States Since 1865*. Baltimore: Johns Hopkins University Press.

Cox, K. L. (2011) *Dreaming of Dixie: How the South Was Created in American Popular Culture*. Chapel Hill: University of North Carolina Press.

Dabney, V. (1978) *Across the Years: Memories of a Virginian*. Garden City, NY: Doubleday, Page.

Daily Times (1889) 'An Old School Southerner', *Daily Times*, Richmond, VA, 4 Dec., p. 8.

Davis, F. (1979) *Yearning for Yesterday: A Sociology of Nostalgia*. New York: Free Press.

De Saussure, Mrs. N. B. (1909) *Old Plantation Days: Being Recollections of Southern Life Before the Civil War*. New York: Duffield and Company.

Eppes, S. B. (1926) *Through Some Eventful Years*. Macon, GA: J. W. Burke.

Gardner, S. E. (2004) *Blood and Irony: Southern White Women's Narratives of the Civil War, 1861–1937*. Chapel Hill: University of North Carolina Press.

Gielow, M. S. (1898) *Mammy's Reminiscences and Other Sketches*. New York: A. S. Barnes.

Gielow, M. S. (1902) *Old Plantation Days*. New York: R. H. Russell.

Gray, R. (1986) *Writing the South: Ideas of an American Region*. Cambridge: Cambridge University Press.

Hale, G. E. (1998) *Making Whiteness: The Culture of Segregation in the South, 1890–1940*. New York: Pantheon.

Hamill, H. M. (1904) *The Old South: A Monograph*. Nashville: Smith and Lamar.

Hobsbawm, E. (1983) 'Introduction: Inventing Traditions', in Hobsbawm, E. and Ranger, T. (eds.) *The Invention of Tradition*. Cambridge: Cambridge University Press, pp. 1–14.

Inscoe, J. C. (2011) *Writing the South Through the Self: Explorations in Southern Autobiography*. Athens: University of Georgia Press.

Janney, C. E. (2013) *Remembering the Civil War: Reunion and the Limits of Reconciliation*. Chapel Hill: University of North Carolina Press.

Litwack, L. F. (1980) *Been in the Storm So Long: The Aftermath of Slavery*. London: Athlone Press.

Logan, F. A. (1950) 'Old South Legend', *Phylon*, 11(3), pp. 234–239.

Mallard, R. Q. (1892) *Plantation Life Before Emancipation*. Richmond: Whittet and Shepperson.

Montgomery, F. A. (1901) *Reminiscences of a Mississippian in Peace and War*. Cincinnati: Robert Clarke Company Press.

Nissenbaum, S. (1996) *The Battle for Christmas*. New York: Vintage.

Pryor, Mrs. R. A. (1909) *My Day: Reminiscences of a Long Life*. New York: Palgrave Macmillan.

Ripley, E. (1912) *Social Life in Old New Orleans: Being Recollections of My Girlhood*. New York: D. Appleton.

Ritterhouse, J. (2006) *Growing Up Jim Crow: How Black and White Southern Children Learned Race*. Chapel Hill: University of North Carolina Press.
Smedes, S. D. (1887) *Memorials of a Southern Planter*. Baltimore: Cushing and Bailey.
Thomas, E. J. (1923) *Memoirs of a Southerner, 1840–1923*. Savannah, GA: s.n.
Williksen, S. and Rapport, N. (2010) *Reveries of Home: Nostalgia, Authenticity and the Performance of Place*. Newcastle upon Tyne: Cambridge Scholars Publishing.
Winston, R. W. (1937) *It's a Far Cry*. New York: Henry Holt.
Wise, J. S. (1899) *The End of an Era*. Boston: Houghton, Mifflin.

14 Only going one way? *Due South*'s role in sustaining Canadian television

Linda Knowles

Introduction

In 1994, a quirky comedy-adventure called *Due South* became the first Canadian television production to air during primetime on a major United States television network. Created by Paul Haggis, who was later to win Academy Awards in two successive years, *Due South* followed the adventures of Royal Canadian Mounted Police Constable Benton Fraser who, to use the explanation given in almost every episode, "first came to Chicago on the trail of the killers of [his] father and, for reasons which don't need exploring at this juncture, . . . remained, attached as liaison to the Canadian consulate." Standing on guard in his red uniform outside the Canadian consulate, Fraser literally represents Canada, as he does in his unlikely partnership with the volatile Chicago police detective Raymond Vecchio, where his strict observance of regulations provides a comic foil to the American's "shoot first, ask questions later" temperament.

The series was one of the highest-rated on a Canadian network; in the US it came second in its time slot, and over 14,000 viewers protested when CBS cancelled it at the end of its first 13-week season. The protest was successful and, despite a chequered career in the US, funding from the UK (BBC), Germany (Pro Sieben Sat 1), and France (TF1) saw it through a further two seasons. It continues in re-runs and, at the time of writing, is being shown in the UK on the True Entertainment channel. Numerous episodes are available on YouTube and the full four seasons are available as DVDs. Though it was by no means a hit on the scale of *Friends, House,* or *Big Brother*, *Due South* represents a major achievement for the Canadian television industry, one which, to date, no other series has repeated. For this reason alone it is worth examining the features which contributed to its success, particularly in relation to questions of authenticity which are raised by the series' cross-border production.

Television only going one way?

Canadian television, as so much else in Canada's culture, is characterised by the particular difficulty it faces in competition with the more powerful industry in the United States. Even before 1952 when the Canadian Broadcasting Corporation

(CBC) began transmitting television programmes, Canadians along the border already owned televisions, their aerials turned due south to catch popular American shows. The Canadian government intended the production of homegrown programmes through the CBC to "protect the nation from excessive commercialization and Americanization," (Beaty and Sullivan, p. 29) but also quickly recognised television's usefulness in fostering national unity by linking communities scattered across vast distances:

> Canada developed a more elaborate and advanced physical structure for delivering radio and television programs than could be found in any comparable country in the world. For example, in 1979 the US had 982 transmitters in operation, but Canada, with a tenth of the population, had 1045 (including rebroadcasting transmitters), a number that grew steadily over the years; by 1981 there were 1225 in operation in Canada, and by early 2010 there were 4918 (including digital).
>
> (Peers, 2010)

Bart Beaty and Rebecca Sullivan's study, *Canadian Television Today* is a useful and comprehensive guide to how this policy worked out in practice. They point out that

> Canadian identity is defined by a fervent desire to be not-American... In this sense, television is made doubly low. First, by its connection to a nation that is seen as the arbiter of all things crass, tacky, and over-blown. Second, by its status as a mass medium that strives for popularity over edification. The goal, for cultural nationalists, has long been to raise the stature of television by ensuring that it is provided with content that is more in keeping with the aesthetic and nationalist values of Canada.
>
> (Beaty and Sullivan, p. 18)

However, Canadian audiences ungratefully persist in preferring the popular to the edifying:

> [V]iewing patterns of Canadians undermine the traditional rhetoric of our cultural distinction... by clearly pointing out [that] we are... a nation that watches *America's Funniest Home Videos* in even greater proportional numbers than Americans.
>
> (Beaty and Sullivan, p. 69)

Indeed, English Canadians (though not French Canadians) are "the only known broadcast market in the world where local dramatic production is not necessarily preferred over foreign content of comparable quality." (Tracey and Redal, quoted in Tate and Allen, 2003, p. 71) As one viewer explained: "Canadians do this thing, which I hate but am sometimes guilty of, where we think that if something is produced in Canada, it can't be any good" (Tate and Allen, 2004, p. 11).

With such a small and reluctant home audience, Canadian production companies found competing with the sheer volume of production from the United States was a financial challenge:

> [A] typical 1974–75 import cost $2000 a half-hour (the actual production cost being roughly $125 000) and yet it could generate a profit of between $20 000 and $24 000 in advertising revenue on the CBC or CTV. Contrast this with domestic production: a half-hour show cost about $30 000 (meaning its production values were inferior to American shows) and realized a profit of $55 on CTV and a loss of $2050 on the CBC. The scheduling of a Canadian show, moreover, usually meant a loss of audiences and revenues in that time slot, since viewers could change channels to find an import that was more appealing.
> (Rutherford, 2006)

Instead, they found their mainstay in cheaply produced syndicated shows like *Star Gate* and *Relic Hunter* designed for the international market and "runaway productions." In other words, "American-funded programs that take advantage of low Canadian currency exchange, cheaper labour, and significant tax incentives to produce their shows here. *X-Files* . . . is the most successful of these" (Beaty and Sullivan, p. 79). The disadvantage with these programmes is that they tend to minimise Canadian content in order to reach an audience presumed to have little or no interest in Canada. Tax incentives and investment opportunities from bodies such as Telefilm Canada (from 1984), while they encouraged domestic programme production, had the disadvantage that they required programmes for international markets to avoid obvious Canadian references (Fremeth, 2010).

> For television the consequence is that
> Between runaway productions and the syndication market, the Canadian television experience can tend to slide into a kind of parlour game in which viewers try to guess the filming locale that is meant to stand in for Chicago, or pick out Canadian actors in bit roles.
> (Beaty and Sullivan, p. 79)

Authentically Canadian in that they are made in Canada and employ Canadians, these productions, however financially viable they may be, do not foster a Canadian television industry able to communicate an authentic sense of Canada either to Canadians or to the world, and do little to further the CBC's mission to stem the tide of Americanisation. As John Doyle points out: "A country is simply inauthentic if its stories are not reflected back to itself" (Bredin, 2012 p. xii). How authentic can Canadian television be if the flow of television culture only goes one way, from its neighbour to the south (Beaty and Sullivan, 2006, p. 89)?

Playing with stereotypes

For all the cultural similarities that enable the United States to dominate Canada's television industry, the two countries have very different experiences of each other. Rightly or wrongly, Canadians assume that they have a more authentic image of the United States than Americans have of Canada. While Canadians are overwhelmed with information about the United States and pride themselves on their knowledge, they are accustomed to ignorance from many Americans who seem to have only a vague understanding of the territory they see as the blank to the north of their weather map from which snowstorms descend. Carol Shields, who was later to make Canada her home, recalled knowing:

> It was cool and quiet there
> with a king and queen
> and people drinking tea
> and being polite and clean
> snow coming down everywhere.
> (Shields, 1992)

Of Winnipeg, she said, "Elsewhere people blink when you say where you're from and half the time they don't know where it is . . . 'You don't mean to tell me you live *north* of North Dakota!'" (Shields, 1993, p. 100). As Hugh MacLennan found, "Boy meets girl in New York, you've got a story; boy meets girl in Winnipeg, who cares?" (Fulford, 2001).

Each country's understanding of the other is also coloured by their differing national myths. Katherine L. Morrison's *Canadians are not Americans: Myths and Literary Traditions* (2003) is an excellent survey of the contrasting principles that distinguish the two nations. In the United States, with a revolutionary history, the principles are (briefly): life, liberty and the pursuit of happiness; making a fresh start; and self-adjusting progress. These are seen by Canadians as recklessness, violence, and abrasiveness or arrogance. For Canada, the principles are: peace, order, and good government; building on the past; and resisting mob rule. These are seen by Americans as timidity, dullness, and deference to authority.

Although *Due South* could be seen as a runaway production, filmed in Toronto to save money on Canadian cast, crew, and locations, it differs from these in having a Canadian character centre stage. Paul Haggis, rather than masking Canadian identity, realised that he could make use of it:

> I could twist around all the archetypes and stereotypes . . . [I] started looking at stereotypes of how Americans view Canadians as really polite but boring people who spend their weekends clubbing seals, and how Canadians viewed Americans as being gunslingers, money grubbing, and imperialistic. And I said, "You know what? I could start a little border skirmish here."
> (Longworth, 2002, p. 167)

Due South took the stereotypically polite and clean Canada and personified it in Benton Fraser, a man "so polite that he keeps giving up his place in a taxi queue . . . ultimately choosing to walk to downtown Chicago, luggage in hand, rather than take up cab space," (Rosenberg, 1994) and partnered him with a Chicago detective with a contrasting personality. Where Fraser is quietly spoken, patient, follows rules scrupulously, and goes unarmed, Raymond Vecchio is loud, abrasive, impatient, and breaks rules with impunity; the quintessential trigger-happy cop.

Paul Haggis clearly intended these two characters to represent their respective countries to comic effect. For example, in the episode entitled 'The Man Who Knew Too Little', Fraser and Vecchio escort a witness across the border into Canada, Vecchio displaying the stereotypical ignorance and indifference of the American towards Canada:

VECCHIO: [to Fraser] You keep your eye on that map. I want a state by state count down until we get to Winnipeg.
FRASER: Windsor.
VECCHIO: Yeah like there's a difference. Damn! I should have brought the snow chains. Do we really got to cross the border? (Series 1 Episode 14)

Windsor is about four hours slightly north east of Chicago, but Vecchio is convinced he's heading to the frozen north (hence the snow tires), and on this occasion, Fraser responds with an un-Fraser-like but very Canadian quip:

FRASER: Yes Ray. Although you know I imagine they'll have a dog sled at the bridge in case we should get stuck. [Laughs, clearly tickled by his own joke]

Initially, *Due South* didn't seem to be the sort of story the Canadian public wanted reflected back at them. They had seen too many fictional Mounties before, from *Rose Marie* to *Sergeant Preston, Dudley Do-Right*, and countless tourist souvenirs. As Paul Gross, who played Benton Fraser pointed out, "Mounties, red coats, sled dogs – to Canadians, it reeks of kitsch" (Conlogue, 1995). Though they were happy to be considered nicer than Americans, the Canadian viewers did not want to be portrayed as naïve or easily gulled, and especially didn't want to be represented by a caricature that looked (in Howard Rosenberg's words) like a "brick of a palooka":

Fraser . . . is a cross between Sgt. Preston and Forrest Gump – so much of a do-gooder wooden automaton that you keep searching his body for a control panel as he mutely stands sentry outside the consulate . . . Stone-faced, square-shooting and single-minded, this guy is the kind of moose who'd walk through a brick wall to get his man.

(Rosenberg, 1994)

Paul Haggis recalled,

I got tons of letters from Canadians . . . And they were *really* objecting to the way I was portraying them as naïve and polite, opening doors for women and such. *These* were the things they were *objecting* to! Schoolteachers would have their entire class write in. They'd say the scene in the airport where he gives away his cab, and he gives away his money to the stranger, is totally unrealistic. And I'd write back and say, "I did that when I came here myself from Canada. I *did* give my money away to strangers. I'm a sucker."

(Longworth, 2002, p. 168)

In the pilot's final scene, Haggis shows Fraser's generosity vindicated as his $100 loan is returned.

As for the Mounties, questions of authenticity were raised even before *Due South* reached the production stage. The RCMP took their image very seriously: according to an 1896 law the RCMP had control over their image in any media in Canada and they refused to give permission for the series to be made. Paul Haggis was initially defiant, regarding potential arrest as good publicity, until he was informed that the licence to manage the RCMP copyright of their image was held by the Disney Corporation.

> Now I would go up against the RCMP, but I didn't want to go up against Disney (*Laughs*.) I mean, that's serious stuff. So I changed the uniform slightly. I had the belt go from left to right rather than right to left across the chest. And I put the badge on the hat (which they don't have) just to make it look a little different.
>
> (Longworth, pp. 167–168)

Haggis deliberately sacrificed authenticity to avoid legal action, but, of course, did nothing to win the approval of the Mounties:

> RCMP spokesman Cpl. Gilles Moreau says that, initially, members of the force criticized Fraser . . . for such technical errors as putting his cross-strap on the wrong shoulder and placing a badge on his Stetson.
>
> (Steele, 1995)

> "[I]t's definitely embarrassing," a real Mountie said . . . The Mountie, requesting anonymity, said that he and a local police officer "howled" with laughter while watching Due South.
>
> (Rosenberg)

There is a certain irony in the thought that the RCMP's anxiety about maintaining an authentic image actually precipitated the errors they objected to in the show. Even the Mounties excused some inaccuracies because "We have to remember this is television" (Rosenberg). In fact, *Due South* treads a fine line between stereotype and authenticity, each one serving to balance the other. For example, *Due South's* pilot episode opens with Benton Fraser taking a dog sled 300 kilometres over a pass in a snowstorm to bring in a man who'd fished over the limit, a clear stereotype of the Mountie who always gets his man. Meanwhile, his colleagues back at the station shake their heads and ask, "A dog sled? Is this guy living in this century?" When it is revealed that the fishing limit is exceeded by tons, it is clear that Fraser's character is an epic exaggeration and is not meant to be taken entirely seriously.

Although Benton Fraser is comically exaggerated – rarely seen with a hair out of place, his red serge uniform always spotless, even when emerging from a dumpster full of garbage – by the end of the pilot episode his unflinching devotion to duty, to the RCMP motto, "Maintiens le droit," harks back to the intention, if not the reality, of the RCMP. Uncovering corruption by his father's killer and arresting a Mountie – "You turned in one of your own" – explain his posting to Chicago, while also revealing inner conflicts that made Fraser's character less stereotypical.

Other elements, such as the appearance of his father's ghost (and later, the ghost of Vecchio senior) as well as car chases, explosions, and derring-do leaps across rooftops made it difficult to decide which genre to assign *Due South*. Lucy Mangan, writing in *The Guardian*, admired it, calling it "a roiling gallimaufry of styles, genres and moods." (Mangan, 2012) This was just what Haggis intended:

> It *was* a little bit of a genre buster because you weren't really sure what the show was. It was a straightforward, cop action show with humor and comedy that, at its best, toyed with deeper and darker themes.
>
> (Longworth, p. 167)

American critics, who compared it to the fish-out-of-water show *McCloud* from the 1970s, were wrong-footed by the lack of an authentic genre in which to place the show. Rosenberg complained, "Mostly, 'Due South' appears uncertain whether to go light or serious, ending in a sort of limbo by awkwardly juxtaposing contradictory tones," while Robin Berkowitz blamed the genre busting for the show's early cancellation by CBS:

> From week to week, CBS viewers could never be sure whether they were in for a frothy action romp or a hard-edged crime-and-punishment saga (usually both). It was the show's greatest creative distinction, and its downfall in terms of finding a mass audience.
>
> (Berkowitz, 1998)

The American audience, however, did not question the authenticity of either Fraser or Vecchio.

> Most American viewers, it seems, thought of Benton Fraser as embodying a vanishing breed of humanity who just happened to come from Canada. It may be that ... they knew that they were not quite as rude and violent as they were portrayed ... Also, most Americans were familiar with the stereotype of the cynical American cop and his the [sic] flexible sense of morality, because it is a type of character that has been seen on American television for decades as has the out-of-town law enforcement official, with a different (and usually more effective) way of doing things.
>
> (Tate and Allen, 2004, pp. 15–16)

If Americans watching the show continued to believe in the clean, polite Canadian (even though many of the villains Fraser confronts are Canadian), Fraser's unvarying courtesy, treating everyone with calm respect, indicates that he is not so much naïve as willing to give others the benefit of the doubt. As Paul Gross says of his character, "So much of TV . . . seems to be dominated by a cynicism and I think there's something really refreshing about somebody that's actually good" (Gross, 1994).

You live *north* of North Dakota?

Despite Canadians' vociferous objections to being characterized as polite and law-abiding, there was no protest at all to the depiction of Benton Fraser as a child of the frozen north (actually born in an igloo as revealed in Series 3 episode 1), perpetuating a myth that Canadians seem unwilling to examine. As Margaret Atwood points out, "Canadians have long . . . invested a large percentage of our feelings about identity and belonging in [the North]" (Atwood 115). Indeed, Canadians tend to rely on the wilderness to give them the distinction that they seem to lack in other areas of their life.

> When traveling in more temperate climates, Canadians take an extravagant pride in telling of temperatures falling to minus forty-five degrees, with the result that tires freeze on parked automobiles and the flat part that rest on the road goes thump-thump as you drive away.
>
> (Atwood 115)

Due South's pilot episode opens in the frozen north and the series concludes with Fraser and Stanley Ray Kowalski on a dogsled heading off to "look for the Hand of Franklin reaching for the Beaufort Sea." Between those two snowy bookends, Fraser doesn't just trade the north for the south; he trades the wilderness for the city. He is warned, "Out there in no man's land, there isn't a better cop in the world. But in Chicago they'd eat you alive in a minute" (*Due South*, "Pilot"). Thus, Chicago becomes a metaphorical wilderness that Fraser must learn to inhabit and for which Toronto provides the filmed location. It is here that *Due South* provides an opportunity to play with an authenticity and a sense of place in the complicated relationship that Canadians have with their cities.

One consequence of Canadians' identification with the wilderness is a tendency to disparage their cities or to fear them as destroyers of nature; an attitude perhaps best summed up in John Robins' 1943 declaration:

> I can approach a solitary tree with pleasure, a cluster of trees with joy, and a forest with rapture; I must approach a solitary man with caution, a group of men with trepidation, and a nation of men with terror.
>
> (Frye (1973) 845)

Cities belong somewhere between trepidation and terror on this continuum, being less than a nation but more than a group of men. Fraser, the ultra-typical

Canadian, is uncomfortable with cities. The biggest place he had ever worked in was Moose Jaw (population 32,973 in 1996), where he only lasted five weeks because he "couldn't adapt to such an urban lifestyle" (*Due South,* "Pilot"). Setting the action in Chicago does more than just give the American audience a point of identification; it turns attention away from the authentic urban experience of most Canadians and maintains the contrast between the Canadian wilderness and the American inner city.

Clean streets and mean streets

In their introduction to *Downtown Canada: Writing Canadian Cities*, Justin Edwards and Douglas Iveson point out that although Canada can be regarded as "one of the most urban countries on earth, with the vast majority of its population concentrated in a handful of cities," Canadians need to be reminded "that 80 per cent of Canadians live in cities," a truism expressed with a "sense of discovery and loss [that] suggests that this is a new and radical shift in Canadian demographic patterns," but which "reveals, in fact, that the city is not yet truly accepted as Canadian" (Edwards and Iveson, 2005, pp. 3–4).

In their desire to be "not-American," the 80 percent of Canadians living in cities are troubled that their cities seem, superficially, to be almost identical to cities in the United States:

> There is a widespread perception that Canadian and American cities are more or less the same. Large North American cities are widely thought of as formless, sprawling entities that eat up valuable farmland and cherished natural features to churn out monotonous subdivisions, malls and office parks.
>
> (Raad and Kenworthy, 1998)

However, there are subtle ways in which Canadian and American cities differ:

> [A]nybody who has spent much time in cities north and south of the 49th parallel will find it hard to ignore the contrast in the liveability of the urban environment in the two countries. Compared to the average US inner city, the inner areas of Canadian cities usually have attractive residential precincts, a strong pedestrian presence in the streets and a vital mixture of shopping, commercial and other activities . . . Their urban fabric is also less fragmented by freeways and parking lots than the average US inner city.
>
> (Raad and Kenworthy)

Chicago and Toronto are actually closely comparable cities, both in population and land area, but Chicago, as portrayed in *Due South*, represents the graffiti-covered, litter-strewn face of the stereotypical American inner city. Though cleanliness in cities is perhaps a relative term, for Toronto's cleaner streets to look authentically like Chicago, walls needed to be graffitied and streets littered, though authenticity was not achieved at once:

"I remember the first day on the set and you could actually read it [the graffiti]," recalls Paul Gross . . . "It said things like 'Go away, we don't like you.'" Since then, the decorators have acquired books of arcane designs by authentic U.S. gang graffiti artists. With a book in one hand and a spray can of washable paint in the other, they correctly "vandalize" Toronto's walls before the cameras roll.

(Wickens, 1995)

It has even been claimed that the graffiti and litter deliberately placed in Toronto streets had to be protected so that they would not be cleaned up before filming could take place (Gross, 1994). For a more comprehensive look at Canada's relationship with cities, see Knowles 2007.

Due South contrasts American and Canadian cities in "Perfect Strangers," (Season 3, Episode 51) in which Fraser and Vecchio travel to Toronto where Vecchio is amazed to see travellers picking up litter at the airport and courteously deferring to each other in the taxi queue. Returning to Chicago, seeing the graffiti and feuding gangs of young people, he resignedly murmurs, "Home again, home again, jiggety jig." Perhaps Vecchio has begun to appreciate that the "brick of a palooka" who gives up his place in a taxi and does not have a problem with being too polite, but with other travellers who are not polite enough.

Despite the warnings, over the course of four seasons, Benton Fraser is not "eaten alive" but, using his wilderness skills and innate courtesy, tames the city, turning the tide of influence so that (at least in his corner of Chicago) it flows from Canada to the US. For example, in Season 1, Episode 18, "The Deal," Vecchio's former school mate (and mafia boss) Frank Zuko equates the graffiti and garbage on the streets of Chicago with a lack of respect and expects Fraser to recognise his protection racket as a necessary evil:

Canadian, right? . . . Well then you do understand. I mean you come from one of those nice clean cities where they have no graffiti, no garbage on the streets, and people treat each other with respect. Right?

Fraser's response is measured and carries a subtle threat. "[I]t's been my experience that many people live their lives thinking that they're respected only to discover they've been merely feared. And fears can be overcome."

The difference between respect and fear that Fraser touches on in this scene can be found in two iconic works of Western fiction: Ralph Connor's *Black Rock* (1898), one of Canada's first best-selling novels, and *The Virginian* (1902), the classic American justification for vigilante rule. In *Black Rock,* "policeman Jackson, her Majesty's sole representative in the Black Rock district" removes a gun from a gambler saying simply, "I'll just take charge of this . . . it might go off," leaving him shocked at "the state of society that would permit such an outrage upon personal liberty" (Connor). In The Virginian, by contrast, the Virginian reacts to being called a "son-of-a –" by taking out his gun and "holding it unaimed . . . . And with a voice as gentle as ever, he issued his orders to the man Trampas: 'When you call me that, SMILE.' " (Wister 1902).

In *Due South*, the equivalent scene becomes a comic misadventure:

FRASER: Excuse me. May I have your attention please? Thank you. Anyone carrying illegal weapons, if you would place them on the bar; you are under arrest.
[Every man in the bar aims a gun at Fraser; a knife is thrown, landing next to his head]
You realize I'm going to have to confiscate that? (*Due South,* "Pilot")

Here, the world of *Black Rock* encounters the world of *The Virginian*, each giving its authentic response to the situation. The ensuing barroom brawl satisfies the American audience's expectation of fictional violence while Fraser's steadfast refusal to use a gun shows that "A man's got to do what a man's got to do," doesn't necessarily mean blowing the other man's head off.

Though he admits envying Vecchio's freedom and spontaneity, Fraser shows the people of Chicago that mutual support, such as standing up to slum landlords and developers, can be more effective than individual efforts (Season 2, Episode 8, "One Good Man, aka Thank You Kindly, Mr. Capra"). In "The Deal," encouraged by Fraser, Vecchio overcomes his fear of Frank Zuko and leaves him humiliated. As "The Deal" ends, Vecchio unloads his gun and locks it away for the night.

Conclusion

In conceiving *Due South* as a "twisting around" of archetypes and stereotypes, Paul Haggis gently mocked the stereotype of the clean, polite, Canadian and the gun-toting, cynical, urban American, giving both Canadian and American viewers something with which they could identify. Employing a "gallimaufry of styles, genres and moods" (Mangan), the series played with the stereotyped images of both countries, overcoming the perceived indifference of the viewers to the south and re-presenting Canada to Canadians and to the world at large. *Due South* proved to Canadians that they did not have to disguise their stories as American but could interest a United States audience in a drama with distinctively Canadian features.

Paul Haggis went on to become internationally famous as an Oscar-winning producer. Paul Gross, who took over from Haggis as *Due South's* producer, appeared in another successful television series, *Slings and Arrows*, as well as producing, directing and starring in two films (*Passchendaele* and *Hyena Road*), described as "two of the most epic Canadian films ever made" (Volmer, 2018), thus establishing his influence on Canadian culture.

While internet streaming and on-demand television have disrupted the old ratings system so that it is difficult to compare levels of popularity today, *Due South's* legacy in Canadian television is clear. Thanks to *Due South*, Canadians no longer regard their television as inevitably worthy but dull. Subsequent Canadian television series, such as the steam-punk historical crime series *The Murdoch Mysteries* and the feminist detective series *The Frankie Blake Mysteries*, employ the quirky humour and knowing nods to Canadian history that characterised *Due South* while the darker crime series *Cardinal* taps into the same dramatic vein that balanced *Due South's* whimsy. All of these, and others, maintain the solid production quality and the air of confidence *Due South* brought to Canadian television, attracting audiences at home and abroad with their unapologetic and authentic Canadian identity.

Bibliography

Beaty, B. and Sullivan, R. (2006) *Canadian Television Today*. Calgary: University of Calgary Press.

Berkowitz, R. (1998) 'Setting a New Course', *Fort Lauderdale: Sun Sentinel*, 6 Jan. Available at: http://articles.sun-sentinel.com/1998-01-06/lifestyle/9801050200_1_detective-ray-vecchio-david-marciano-benton-fraser [Accessed 15 Sept. 2011].

Bredin, M. *et al*. (2012) *Canadian Television: Text and Context*. Waterloo: Wilfrid Laurier University Press.

Brooke, R. (1916) *Letters from America*. London: Sidgwick and Jackson.

Buendgens-Kosten, J. (2014) 'Authenticity', *ELT Journal*, Oct., 68(4), pp. 457–459.

Conlogue, R. (1995) 'Profile: Giving Gross His "Due": CBS' Canadian Mountie Is Perfectly Suited for Absurdly Do-Right Role', *L. A. Times*, 8 Jan. Available at: www.latimes.com/archives/la-xpm-1995-01-08-tv-17436-story.html [Accessed 018082019].

Connor, R. (1898) *Black Rock: A Tale of the Selkirks*. Toronto: Westminster. Available at: www.gutenberg.org/files/3245/3245-h/3245-h.htm [Accessed 14 Aug. 2019].

Edwards, J. D. and Ivison, D. (2005) *Downtown Canada: Writing Canadian Cities*. Toronto: University of Toronto Press.

Fremeth, H. (2010) *Television*. Available at: www.thecanadianencyclopedia.ca/en/article/television [Accessed 18 Aug. 2019].

Frye, N. 'Conclusion', in Klinck, C. F. (1973) *Literary History of Canada*. Toronto: University of Toronto Press, p. 845.

Fulford, R. (2001) 'Column', *The National Post*, 6 Jun. Available at: www.robertfulford.com/CanadianNovelists.html [Accessed 19 Aug. 2019].

Gross, P. (1994) 'Interview', *This Morning*. CBS. 22 Sept., Season 2 DVD.

Gross, P. and Carney, R. B. (1999) 'Call of the Wild, Part 2', *Due South*. Available at: www.youtube.com/watch?v=uPWe6bFy4AQ [Accessed 18 Aug. 2019].

Knowles, L. (2006/2007) 'Kronk City? Canadian Cities in the Novels of Carol Shields', *London Journal of Canadian Studies*, 22, pp. 85–114.

Longworth, J. J. (2002) *TV Creators: Conversations with America's Top Producers of Television Drama*. Syracuse: Syracuse University Press.

Mangan, L. (2012) 'Your Next Box Set: Due South', *The Guardian*, 13 Mar. Available at: www.theguardian.com/tv-and-radio/2016/jul/27/tvs-most-underrated-shows-pulling-top-gear-time-trumpet [Accessed 14 Aug. 2019].

Morrison, K. (2003) *Canadians Are Not Americans*. Toronto: Second Story Press.

Peers, F. W. and Harada, S. (2010) *Radio and Television Broadcasting*. Available at: www.thecanadianencyclopedia.ca/article/radio-and-television-broadcasting [Accessed 26 Apr. 2019].

Raad, T. and Kenworthy, J. (1998) 'The US and Us', *Alternatives Journal*, Winter. Available at: www.questia.com/magazine/1G1-20354687/the-us-and-us [Accessed 18 Aug. 2019].

Rosenberg, H. (1994) 'When a Mountie's as Thick as a Brick', *Los Angeles Times*, 26 Sept. Available at: www.latimes.com/archives/la-xpm-1994-09-26-ca-43258-story.html [Accessed 14 Aug. 2019].

Rutherford, Paul (2006) *Television Programming*. Available at: www.thecanadianencyclopedia.ca/en/article/television-programming [Accessed 14 Aug. 2019].

Shields, C. (1992) *Coming to Canada*. Ottawa: Carleton University Press.

Shields, C. (1993) *The Republic of Love*. London: Harper Collins, Flamingo.

Steele, S. (1995) 'Opening Notes', *Maclean's*, 6 Feb. Available at: https://archive.macleans.ca/article/1995/2/6/opening-notes#!&pid=10 [Accessed 14 Aug. 2019].

Tate, M. A. and Allen, V. (2003) 'Integrating Distinctively Canadian Elements into Television Drama: A Formula for Success or Failure? The Due South Experience', *Canadian Journal of Communication*, 28(1), pp. 67–83.

Tate, M. A. and Allen, V. (2004) *Due South and the Canadian Image: Three Perspectives*. Available at: https://mtateresearch.com/DSARCSppr.pdf [Accessed 6 Feb. 2019].

Volmer, E. (2018) *Paul Gross Plays a Very Different Mountie in CBC's Gritty Caught*. Available at: https://calgaryherald.com/entertainment/television/paul-gross-plays-a-very-different-mountie-in-cbcs-gritty-caught [Accessed 14 Aug. 2019].

Wickens, B. (1995) 'Opening Notes: American Graffiti Canadian-Style', *Maclean's*, 12 Nov. Available at: https://archive.macleans.ca/article/1995/11/27/opening-notes [Accessed 14 Aug. 2019].

Wister, O. (1902) *The Virginian, a Horseman of the Plains*. New York: Palgrave Macmillan. Available at: www.gutenberg.org/files/1298/1298-h/1298-h.htm#link2H_4_0005 [Accessed 14 Aug. 2019].

Index

Age of Reason 163
Americana 61–74
American Indians 75–92, 95–111
American sublime, the 21, 129–30, 132–3, 136
ancestry 118
anti-hero 127–8
Austin 12–16
authenticity: authenticity paradox 112–20; staged authenticity 2, 97–8, 102, 106

Barbarians are Coming, The 3, 46–8, 55
Beach Boys 41–53
belonging 134–7
Bordentown, New Jersey 174–6

capitalism: friendly 17
Carr, Emily 99
Chicago 108, 131, 148, 192–202
childhood 128, 134, 184–90
Christmas 180, 189
citizenship 31, 83–4, 92, 170
Civil War 6, 9–10
cognitive dissonance 118–19
comedy drama 192, 198
comic books 15–16, 19
concept towns 146, 149, 155
Confederacy 189
conservative 64, 126, 156, 170, 173
country music 62–74
cowboy 65, 123–4, 127, 129
Crisis Papers, The 162
culinary 30–40

decolonialism 5, 96–7, 107–8
Diné 80–4, 90, 92–3
Due South 191–204
dystopia 22, 25, 103

fake: news 1–2
folklore 6, 100, 107, 147
frontier myth 1, 4, 9, 47, 62–72, 98, 113, 116, 120

Gilpin, Laura 4, 75–83

Haida: Haida Gwaii 4–5, 96–108
hauntology 3, 21
hegemony 10
Helldorado Festival 123–33
heterotopia 20, 22, 24, 28, 71–3, 75
hyperreality 1, 5, 23, 25; magical hyperreality 96–111
hypertextuality 22

immigration 33–40, 69, 164; Norwegian 145, 147–53, 158; Swedish 135, 137–9, 141, 145
incompatibility 19
invented tradition 31, 35–7, 39, 181

Jim Crow 180, 183, 188, 189

landscape: cinematic 23, 28, 114–18, 120
liberalism 15, 17, 25, 65, 67
life-writing 180–91
Lost Cause 183
Louie, David Wong 30–40

magic: magical hyperreality (*see* hyperreality); magic realism 97, 99–102, 104–5
Manifest Destiny 10, 114, 117, 120
marginalization 35, 61, 62
mediated space 19–20, 22, 28, 113
memory: collective memory 31; memorialization 6, 161, 163, 168, 173, 176

military service 9, 30, 82–4, 86, 126, 171
mobility 4, 69–72
monuments 6, 21–2, 161–76
Morristown, New Jersey 162, 170–2
Mount Horeb 6, 145–58

Navajo 4, 75–92
negative capability 113–14
New Rochelle, New York 162, 165–6, 168, 170, 172
nostalgia 2, 5–6, 34, 38, 91, 98, 120, 124, 135–7, 152, 180–9

Old South 180
othering 9, 39, 99, 104–5; exotic 31–2, 35

Pacific Northwest *see* Haida, Haida Gwaii
patriotism 82, 84, 172
Pet Sounds *see* Beach Boys
photography 19, 64, 75–91, 99, 113–14, 128, 188
plantation 6, 180–9
Portland 8, 13–14, 16
postbellum 2, 6, 180–9
poverty 62–3, 184
psycho-geography 22, 25

queer studies 4, 75

race: appropriation 3, 5, 36, 81, 97
radicalism 3, 6, 16, 20, 35, 66
Reagan, Ronald 173
reenactment 124–8, 132
resilience 65
Roosevelt, Franklin D. 169–70

Royal Canadian Mounted Police 192
rurality 61–74, 134–44

Santa Cruz 8, 13–17
simulacrum 19–20, 22–7, 96–7, 100, 104–5, 108
slavery 164, 175, 180, 181–9
slogans 8, 12, 15–18, 25
small-town America 4, 61–3, 65, 74
Southern California 41–7, 54
sovereignty 4, 77, 79, 81, 84, 86
spectacle 3, 5, 24, 114, 114, 120
Springsteen 2
staged backstage 1, 97, 106
story maps 118
strange 16, 19, 31
surf culture 13, 41
Surfin' U.S.A. 42
Super Natural British Columbia 4, 96, 100–2, 104, 108

Thomas Paine 161–76; Thomas Paine National Historical Association 167–8
Tombstone 5, 123–32
Toronto 195, 199–201
trolls 6, 145–6, 148, 154–8

UNESCO 102, 107
utopia 22, 24, 173

Vlautin, Willy (Richmond Fontaine) 4, 61, 65–72

weaving 79–81, 83, 85–7, 90, 92
weird 3, 8, 9, 11–17
Wilson, Brian 41–54